AI Governance and IT Risk Management

Overview of Process and Needs for Governing and Artificial Intelligence Environment

Claude Louis-Charles, Phd

Cybersoft Publishing LLC

Fort Washington, MD 20744

Drclaude.net

Second Edition April 2026

1 Introduction

A regional hospital network rolled out an AI triage tool with little oversight after passing basic checks. Fourteen months later, an audit revealed the model consistently underrated severity for some patients based on age, insurance, and chief complaint. Lacking governance measures like risk assessment and bias testing, the issue went unnoticed. Legal and compliance teams then spent eight months recreating missing documentation that should have been established before deployment.

That scenario is not exceptional. It is representative of what happens when capable technology outpaces the organizational structures designed to manage it. AI systems are now embedded in financial decisions, hiring workflows, supply chain operations, customer communications, and patient care. The managers responsible for those functions inherit accountability for outcomes they often had no hand in designing. The governance gap between what AI can do and what organizations have in place to oversee it is the central problem this book addresses.

This chapter introduces the urgency of governance, outlines key risk categories for AI programs, describes the evolving AI landscape, and defines the scope and method of subsequent chapters. It emphasizes that managers should actively oversee AI systems rather than passively accept their outputs.

Diagram 1.1: The Governance Gap

The widening disparity between accelerating AI capability and lagging organizational governance structures requires proactive strategic action.

1.1 Why AI Governance Matters Now

Three forces have converged to make AI governance an immediate operational concern rather than a future aspiration. The first is the pace of deployment. Organizations that spent years piloting narrow AI tools have, over the past three years, moved to broad-based deployment of systems that touch core processes. The transition from experiment to production at scale happened faster than governance frameworks could mature. What was once a data science project now routes loan applications, screens resumes, flags fraud, or schedules field technicians. Each of those functions carries regulatory exposure, workforce impact, and reputational risk that were not fully considered when the system was approved.

The second force is regulatory momentum. Legislatures and regulators in the European Union, the

United States, Canada, and a growing number of other jurisdictions have moved from issuing principles-based guidance to codifying concrete requirements. The EU AI Act imposes conformity assessments, technical documentation mandates, and post-market monitoring obligations on high-risk applications. Several U.S. states have enacted algorithmic accountability laws requiring impact assessments and human oversight mechanisms. Even where specific AI statutes do not yet exist, existing privacy, financial services, healthcare, and employment law frameworks have been applied to AI-driven decisions. Managers who treat governance as a future compliance exercise are already behind the regulatory curve.

Organizational experience shows that AI risks are real, including failures such as biased credit models and harmful chatbots. The focus has shifted from debating AI governance to figuring out how to build programs that work within budget limits, shifting priorities, dynamic vendor ties, and engineering teams that see governance as friction.

Together, these forces define the imperative of governance. A manager who understands AI capabilities but lacks a framework for overseeing their deployment is exposed. So is the organization. The goal of this book is to close that exposure through practical, enforceable governance that integrates with existing IT and compliance functions rather than adding a parallel bureaucratic layer. Every chapter is oriented toward that integration: connecting governance requirements to the management workflows, risk functions, and operational disciplines that organizations already maintain.

It is also worth being precise about what governance is not. Governance is not a brake on AI adoption. Organizations that govern AI responsibly can and do move quickly. Still, they move quickly on a foundation of documented decisions, tested systems, and assigned accountability that protects them when something goes wrong. The alternative, moving quickly without that foundation, is not agility but exposure. The hospital network in the opening scenario did not save time by skipping governance; it deferred the cost and compounded it.

Diagram 1.2: Three Forces Driving Governance Urgency

The convergence of rapid deployment, increasing regulation, and evolving organizational experience makes immediate governance establishment critical.

1.2 The Evolving AI Landscape

Understanding what is being governed requires a working map of current AI capabilities and deployment patterns. The manager does not need to understand backpropagation or transformer architecture in depth. But a manager who cannot distinguish between a rules-based automated decision system, a supervised machine learning model, and a large language model will struggle to apply the right governance controls to each. These are different technologies with distinct risk profiles, failure modes, and oversight requirements.

Rules-based automation, which uses clear, predictable logic to process structured inputs, remains the most widely used form of AI-related technology in businesses. These systems are easy to monitor and regulate because their reasoning can be reviewed and their results linked to specific conditions. Managing these systems mainly means ensuring the rules match current policies, that any exceptions are handled through formal escalation procedures rather than unofficial workarounds, and that outdated or biased criteria are not embedded in the rules. Over time, rules-based systems can become less accurate as original conditions change, and maintenance often becomes less of a priority once the system is up and running.

Supervised machine learning models learn statistical relationships from historical data and apply those relationships to new inputs. Loan underwriting models, churn prediction systems, image classification tools, and fraud detection engines fall into this category. Governance requirements here are more complex: the training data must be representative and free of inappropriate proxies, the model's behavior must be validated before deployment, performance metrics must be monitored over time, and the model must be retrained or replaced when its operating environment drifts away from the conditions it was trained on. Drift, the degradation of model accuracy as real-world data distributions shift away from the training distribution, is among the most common causes of AI system failures in production. A model that was highly accurate at deployment may become significantly less reliable within months without ongoing monitoring.

Large language models and generative AI systems represent the newest and most rapidly expanding category. These models produce open-ended outputs, text, code, images, and synthetic data in response to prompts that are often informal and unpredictable. Their failure modes include hallucination, prompt injection, generation of harmful content, and the exposure of sensitive information present in training data. They require governance controls that did not exist in most IT organizations as recently as two years ago: prompt management policies, output-review workflows, content-safety filters, and vendor contractual protections that specify how proprietary data will be used in model training. Generative AI has also introduced new categories of workforce risk; automation of professional judgment functions that were previously considered beyond the reach of AI, which carry governance implications beyond the technical.

Most large enterprises now operate all three categories simultaneously, often within the same business process. A customer service workflow might use rules-based routing to classify the inquiry, a machine learning model to predict customer intent, and a large language model to draft the response. Governance must follow the same path, applying appropriate controls at each stage rather than treating AI as a monolithic category to be addressed with a single policy and a single review process.

Diagram 1.3: Three Categories of AI in the Enterprise

Rules-Based Automation

Rules automation

Key governance requirements:
- Rules-based automation and system design
- Strict adherence to procedures

Supervised Machine Learning

Supervised — Machine learning

AI

Key governance requirements:
- Model training and data quality
- Continuous system monitoring

Generative AI

Generative technology

Key governance requirements:
- Content oversight and ethics
- Responsible content creation

Three key functional categories of AI technology that inform different enterprise strategies.

1.3 Primary Risk Categories

Effective governance begins with a clear-eyed view of what can go wrong. The risk landscape for AI systems is broader than many managers initially recognize. It encompasses not only the obvious failure modes that result in incorrect outputs, but also systemic risks arising from how AI is integrated into organizational decision-making, workforce structures, and external relationships.

Bias and fairness risk arise when an AI system produces outcomes that treat individuals or groups differently based on characteristics that should not influence the decision. This can occur because training data reflects historical inequities, because proxy variables substitute for protected characteristics, or because the model was optimized on metrics that inadvertently disadvantage certain populations. The consequences include regulatory violations, reputational harm, and, most concretely, demonstrable damage to real people: denied benefits, mispriced services, unfair treatment in hiring or lending. The insidious quality of bias risk is that it is often invisible to the deploying organization, while its effects are clearly visible in affected communities.

Privacy and data risk encompass unauthorized processing of personal data, retention beyond permitted periods, cross-context use of data in ways that violate consent, and the exposure of sensitive information through model outputs. AI systems that are trained on personal data, ingest personal data at inference time, or generate outputs referencing personal information are all subject to privacy obligations under the GDPR, CCPA, HIPAA, and similar frameworks. Managers must understand where personal data enters and exits the AI system and ensure that data-handling practices align with applicable law and the organization's published privacy commitments.

Operational and safety risk covers the possibility that an AI system will fail in ways that disrupt business operations, harm customers or employees, or cause physical harm in contexts where AI drives real-world actions. This includes model degradation over time as data distributions shift, adversarial inputs designed to manipulate model behavior, and cascading failures in systems where AI outputs feed into downstream automated processes. Safety risk is highest in domains where AI decisions have physical consequences, such as autonomous logistics, clinical decision support, infrastructure management, and financial market operations. In these domains, the consequences of AI failure are not limited to business disruption; they can include loss of life, significant physical harm, or systemic financial damage.

Governance and accountability risk is the meta-risk: the organization lacks the structures, documentation, and processes needed to demonstrate responsible AI use to regulators, auditors, and the public. Even a technically sound AI system creates significant liability exposure if the organization cannot show that appropriate oversight was applied, that risks were assessed, and that incidents were handled responsibly. This risk is fully within the manager's control. Technical risk requires engineering expertise to address; governance and accountability risk requires organizational discipline, management commitment, and the program-building skills that this book develops.

Diagram 1.4: The Four Primary AI Risk Categories

1.4 Scope and Approach of This Book

This book covers AI governance and IT risk management as integrated disciplines. They are separated in some organizational structures; data science teams own AI, IT owns risk management, but that separation is an organizational accident rather than a principled design. AI systems are IT systems. They carry data, operational, vendor, and compliance risks. Managing them responsibly requires both AI-specific governance knowledge and fluency in established IT risk management frameworks. This book provides both.

The scope is the enterprise manager: someone responsible for an IT function, a business unit with AI-enabled processes, or an enterprise-wide AI or data program. The book does not assume deep technical expertise in machine learning or software engineering. It does assume that the reader understands organizational dynamics, has managed IT projects or operations, and is accountable for results in a regulated or compliance-sensitive environment. The frameworks and checklists in each chapter are designed to be actionable by someone who manages the use of AI systems rather than someone who builds them.

The approach throughout is practical and decision-oriented. Each chapter identifies the decisions a manager must make, the artifacts that must exist, and the behaviors that distinguish a functional governance program from a paper exercise. Where frameworks and models are introduced, they are presented as tools for action rather than abstract taxonomies. The goal is not to have readers describe AI governance, but to have them practice it, so they can walk into a governance review, a regulatory examination, or an incident response with the knowledge and documentation to demonstrate responsible stewardship.

The book intentionally avoids two failure modes commonly discussed in the governance literature. The first is vendor-centric analysis: treating AI governance primarily as a procurement or vendor-assessment problem rather than as an ongoing organizational capability. Vendors provide systems; organizations are responsible for how those systems are deployed and what outcomes they produce. Vendor due diligence is one component of a governance program, not a substitute for one. The second failure mode is regulatory-compliance-only framing: treating governance as the minimum required to avoid regulatory sanction. That framing produces brittle, checkbox-driven programs that pass audits and fail in practice. The standard in this book is genuine accountability: governance structures that would catch real problems, not just satisfy auditor inquiries.

1.5 How to Use This Book

The chapters that follow are organized to move from the conceptual to the operational. Chapters on the social and ethical dimensions of AI (Chapter 2) and on governance structures (Chapter 3) establish the "why" and the "who". The legal framework chapter (Chapter 4) provides the regulatory context that shapes what governance must accomplish. Implementation chapters address how to build programs, draft policies, manage the AI development lifecycle, and operationalize data privacy. The final chapters address audit and assurance, specialized governance for generative AI, and playbook artifacts ready for immediate use.

A manager building a governance program from scratch should read the book sequentially, as each chapter assumes familiarity with earlier concepts. A manager who is upgrading a partial program can use the chapter structure as a diagnostic: identify where the organization's current practices are weakest and focus there first, then return to adjacent chapters to close gaps. Each chapter's manager's checklist makes this targeted reading practical — the checklists are designed to be usable as standalone audit instruments even by readers who have not worked through the full chapter.

Practitioners preparing for a regulatory review or audit will find the legal framework and assurance chapters most useful. Managers evaluating a specific AI system or vendor should combine the AI development lifecycle chapter with the third-party risk policy module in the IT policy framework. Those preparing organizational governance structures should combine the governance committee and decision rights material in Chapter 3 with the implementation roadmap in Chapter 5.

The manager's checklists at the end of each chapter are intended as working documents. They should be completed for each AI system within scope, with evidence references where applicable. A completed checklist is not a substitute for the full governance record. Still, it provides a rapid-assessment view that is useful for periodic program reviews, for briefing senior leadership, and for identifying gaps that require attention before a formal audit or regulatory inquiry.

Diagram 1.5: Book Navigation Map

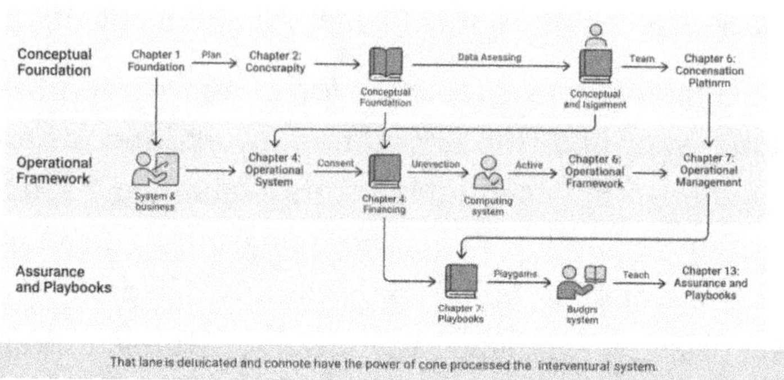

1.6 The Manager's Governance Role

The manager's role in AI governance is distinct from the roles of the AI engineer, the data scientist, the legal counsel, and the compliance officer, though it intersects with all of them. The manager is the integrator: the person who assembles the inputs from technical, legal, and compliance functions into a coherent program; who allocates the resources that governance requires; who makes the judgment calls when technical, ethical, and business considerations pull in different directions; and who owns the outcomes.

That ownership is not metaphorical. When an AI system produces discriminatory outcomes, the manager of the function that deployed it bears accountability to the organization's leadership, to regulators, and in some jurisdictions, to the individuals harmed. When an AI system fails operationally, the manager responsible for the business process it supports is accountable for continuity. When an AI vendor mishandles data, the manager who executed the vendor relationship owns the incident response. Governance is the mechanism through which managers exercise accountability proactively rather than defensively.

This means the governance role is not a secondary responsibility to be delegated to a compliance analyst or handled through periodic policy reviews. It requires ongoing attention to the AI systems within the manager's scope: knowing what they do, how their outputs are used in decisions, what their performance metrics show, and whether the conditions under which they were originally approved have changed. This does not require the manager to become a machine learning practitioner. It requires the manager to develop fluency in a set of governance questions that can be asked of any AI system, regardless of its technical architecture.

The governance role also requires the manager to maintain organizational conditions that enable governance to function. That means protecting the time and resources required for risk assessment, documentation, and monitoring against the constant organizational pressure to prioritize deployment velocity. It means supporting the escalation culture that makes it safe for engineers and analysts to overcome surface concerns. It means ensuring that vendors are held to governance standards through contractual mechanisms, not just informal expectations. And it means communicating with senior leadership and the board in terms that connect AI governance to the business risks and strategic objectives they are responsible for.

The manager who fulfills this role is not performing a compliance function; they are exercising professional judgment in a domain that has become central to organizational performance and risk. The chapters that follow provide the frameworks and tools for that judgment. The most important contribution this book can make is not to provide a template that managers fill in but to develop the analytical habits that allow managers to ask the right questions in situations the templates did not anticipate.

Diagram 1.6: The Manager as Governance Integrator

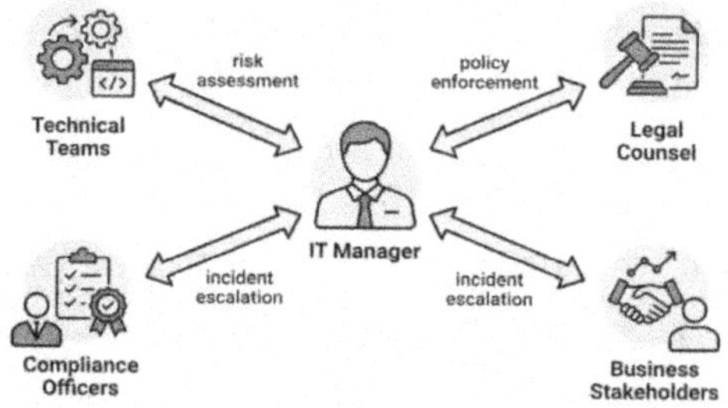

This chart is inclusive, the manager as system IT manager = lead team fontrolling the governance comprence.

1.7 Key Terms Defined

Governance is the system of structures, processes, and accountabilities through which an organization manages its AI systems, deciding what is built, approving what is deployed, monitoring what is operating, and responding when something goes wrong. Governance differs from management in that it operates at the policy and oversight levels rather than at the day-to-day execution level. A manager who approves deployment is performing a governance function. A manager who checks whether a model is performing within acceptable parameters is performing a governance function. Both are distinct from the engineering work of building or operating the model.

Risk management is the discipline of identifying, assessing, treating, and monitoring risks. In the AI context, risk management addresses the probability and consequences of AI system failures, adverse outcomes, and compliance violations. It includes both proactive control actions taken before deployment to reduce the likelihood of harm and reactive controls, incident response, root cause analysis, and remediation after harm occurs. Risk management without governance is an analytical exercise without organizational authority. Governance without risk management is a process without substance.

A control is any action, process, policy, or technical measure designed to reduce a specific risk. Controls can be preventive, they stop an adverse event from occurring; detective, they identify an adverse event after it has occurred; or corrective, they restore normal operations or conditions after an adverse event. A mature governance program deploys all three types. Preventive controls are the most valuable but cannot catch every failure; detective controls are necessary because preventive controls are imperfect; corrective controls are essential because even well-governed systems fail.

A model card is a structured documentation artifact that describes an AI model's purpose, training data sources, performance metrics across relevant subgroups, known limitations, and intended use conditions. Model cards are both a governance artifact and an operational communication tool — they help people who use or oversee a model understand what it was built to do and where it should not be trusted. An organization that cannot produce a model card for a deployed system cannot demonstrate that the deployment was governed responsibly.

A risk register is a catalog of identified risks, each characterized by its likelihood, potential impact, current control state, and assigned owner. In AI governance, the risk register tracks both system-level risks specific to AI systems and their known failure modes, and program-level risks, gaps in the governance program itself that leave the organization exposed. The risk register is the primary instrument through which governance stays current: risks that have been remediated are closed, new risks are added, and the register as a whole provides a real-time view of the organization's AI risk exposure.

Human-in-the-loop refers to governance and operational designs in which a human reviewer is present at critical decision points, with the authority and practical ability to override, delay, or escalate an AI-driven recommendation. The definition matters: a human who receives AI recommendations and is never trained or empowered to question them is not meaningfully in the loop. Human-in-the-loop requirements are most significant in high-stakes domains and are increasingly codified in regulation. The governance obligation is not merely to insert a human step in the process but to ensure that the human at that step has the information, training, and authority to exercise genuine judgment.

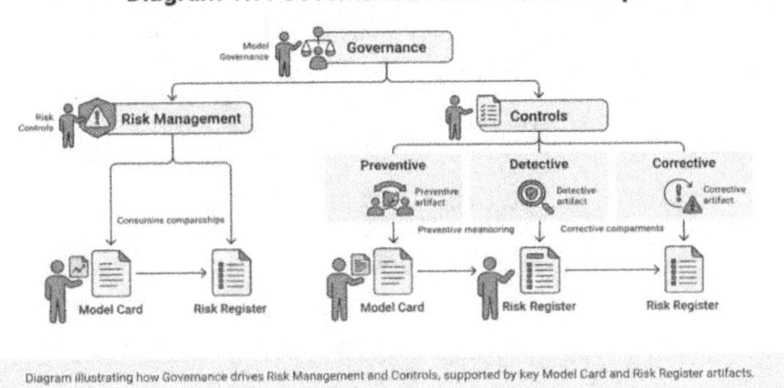

Diagram 1.7: Governance Term Relationships

Diagram illustrating how Governance drives Risk Management and Controls, supported by key Model Card and Risk Register artifacts.

1.8 Manager's Checklist

The following checklist summarizes the minimum behaviors a manager should exhibit before considering an AI governance program operational at its foundational level. Each item corresponds to a concept introduced in this chapter.

Inventory: Maintain a current inventory of all AI systems within your scope, including vendor-provided systems, internally developed models, and AI features embedded in enterprise software platforms. Each entry should identify the system's purpose, the decision or process it supports, the data it processes, and the business owner. An undocumented AI system is an unmanaged liability.

Risk categorization: Assign each AI system a risk tier based on the potential severity of its failure modes, the sensitivity of the data it processes, and the regulatory requirements that apply to its domain. High-risk systems require more intensive governance controls than low-risk systems. The tier assignment should be documented and reviewed when material changes occur to the system or its operating context.

Accountability assignment: Every AI system in your scope should have a named owner, a specific person, not a team, who is responsible for its performance and compliance. That owner should have both the authority and the resources to act on governance requirements. Accountability that cannot be exercised is accountability in name only.

Baseline documentation: Before or immediately after deployment, each AI system should have a model card or an equivalent artifact that describes its purpose, data sources, performance characteristics, and known limitations. The absence of this documentation is a governance gap, not a compliance technicality. It means the organization cannot demonstrate that the deployment decision was made with adequate information.

Monitoring commitment: AI systems in production must be monitored. Performance metrics, fairness indicators, and operational health signals should be reviewed on a defined schedule. The schedule should be risk-appropriate, and high-risk systems require more frequent review. Monitoring that exists only on paper, scheduled but not conducted, is a governance failure that will be exposed in an audit or incident investigation.

Incident readiness: Ensure that the people responsible for AI systems know how to identify and report a potential incident, and that there is a defined escalation path from initial detection to organizational response. Discovering the escalation path during an incident is too late. Incident readiness requires periodic testing: table-top exercises or structured scenario reviews that verify the path works before it is needed.

Executive visibility: Senior leadership and the board should receive regular reporting on the organization's AI risk posture. If no mechanism exists for AI governance information to reach executive and board-level audiences, that mechanism must be created. AI governance that is confined to the operational level cannot sustain the organizational attention and resource allocation it requires.

1.9 What This Chapter Establishes

AI governance is not a compliance exercise layered on top of technical work. It is the organizational infrastructure that enables the responsible deployment of AI. The pace of AI deployment, the momentum of AI-specific regulation, and the accumulating evidence of AI failures in practice have made governance a present operational requirement for every enterprise that uses AI in its core processes.

The manager is the linchpin of that infrastructure. Technical teams can build reliable systems; legal and compliance teams can track regulatory requirements; auditors can assess program quality. But only the manager who owns the business process can integrate those inputs into a governance program that functions under real organizational conditions that survives budget pressures, personnel changes, and the constant pull toward deployment velocity that characterizes enterprise AI programs.

The chapters that follow provide the frameworks, templates, and decision guidance to make that possible. They are organized to build progressively: social and ethical foundations, governance structures, legal requirements, implementation, policy, lifecycle controls, privacy, regulatory frameworks, generative AI governance, and assurance. Each chapter contributes to a program that is not merely documented but also functional, catching real problems, responding to real incidents, and demonstrating genuine accountability to the auditors, regulators, and communities that will eventually ask.

2 AI as a Social Paradigm

A large municipal transit authority deployed a predictive scheduling system to optimize bus routes and staffing levels. The system was technically impressive: it reduced cost-per-mile, improved on-time performance metrics, and satisfied procurement requirements without exceeding budget. Within eighteen months, rider advocacy groups published a detailed analysis showing that route optimization had systematically reduced service frequency in lower-income neighborhoods while increasing it in commuter corridors serving wealthier districts. The transit authority had treated the system as an operations tool. Advocacy groups and city council members treated it as a policy decision. Both were right. The authority had no framework for evaluating the social implications of algorithmic resource allocation before it became a public controversy, a civil rights complaint, and a mayoral priority.

This chapter examines the gap between treating AI as a technical optimization exercise and understanding it as a social intervention. Every AI system that makes or influences decisions affecting people operates in a social context. The manager who ignores that context is not managing the system's full risk. The manager who engages with it is practicing governance at the level of maturity that regulators, community stakeholders, and organizational leadership increasingly expect.

This chapter examines AI's societal footprint, the ethical imperatives that governance programs must operationalize, the practice of stakeholder mapping, frameworks for social risk assessment, equity and fairness considerations, community expectations, and the specific accountability that AI governance places on managers. The goal is not to turn managers into ethicists but to give them the analytical tools to ask the right questions before a system is deployed and to recognize when the answers require a change of direction.

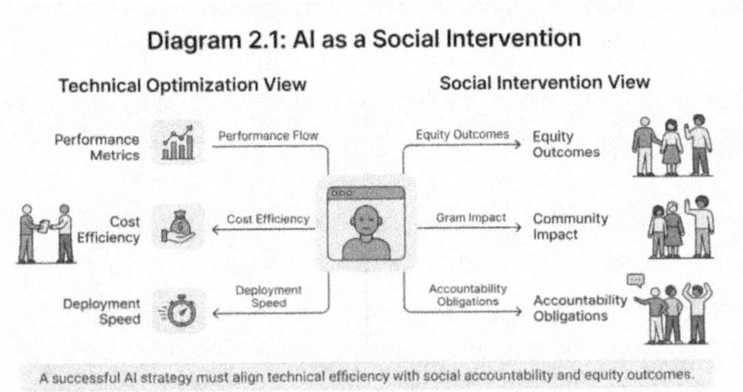

2.1 AI's Societal Footprint

AI systems have a societal footprint, a set of effects on communities, institutions, and power relationships that extends beyond the immediate organizational goal the system was designed to serve. Recognizing that footprint is a precondition for governing it. A manager who can only see the system's performance metrics cannot see the full risk or anticipate the challenges that will eventually arise outside the organization's technical boundaries.

The footprint operates at multiple levels. At the individual level, AI-driven decisions affect real people's access to credit, employment, healthcare, housing, benefits, and public services. When those decisions are wrong, when the model misclassifies, underscores, or applies inappropriate criteria, the person affected may have no visible mechanism to identify the cause or seek correction. The opacity that makes AI systems powerful also makes their errors difficult to surface and contest. This is the source of the accountability requirement: organizations that substitute algorithmic decisions for human judgment must maintain the ability to explain, audit, and correct those decisions.

At the institutional level, AI systems adopted across an industry or sector can reshape its practices and norms. Credit scoring models that penalize characteristics correlated with race have produced racially stratified credit access at scale. Hiring algorithms trained on historical workforce data have perpetuated historical exclusion. The individual deployment decision appears neutral, optimized for predictive accuracy, but the aggregate effect is institutional. A manager governing a single AI system within a single organization cannot fully control that institutional dynamic, but understanding it shapes how the system is designed and what safeguards are required.

At the public trust level, the manner in which organizations deploy AI affects the legitimacy of AI as a tool of governance and service delivery. Communities that experience AI-driven decisions as opaque, arbitrary, or discriminatory develop justified skepticism toward institutions that use those systems. That skepticism is a governance risk: it generates regulatory pressure, litigation exposure, and operational friction when affected communities organize to resist or challenge AI-driven outcomes. Proactive social risk management reduces that friction by building the transparency and accountability that sustain trust over time.

The practical implication for managers is that AI impact assessment must include a social dimension. Before a system is deployed, someone must ask: Who is affected by this system's outputs? How does the system's behavior vary across different populations? What recourse is available to people who believe the system has treated them unfairly? Are there communities or groups for whom this system creates disproportionate risk or burden? Those are governance questions, not technical ones. The technical team can provide data to answer them, but the manager must ensure they are asked and that the answers are reflected in deployment decisions.

Organizations that have built social impact assessment into their governance workflows report that the process surfaces concerns earlier, when they are cheaper to address. A bias pattern identified during pre-deployment testing requires a model adjustment. The same pattern identified eighteen months into production may require a full program review, regulatory disclosure, and a public response. The economics of social governance strongly favor frontloading: the cost of asking the right questions before deployment is a fraction of the cost of answering them under regulatory pressure afterward.

Diagram 2.2: Levels of AI Societal Footprint

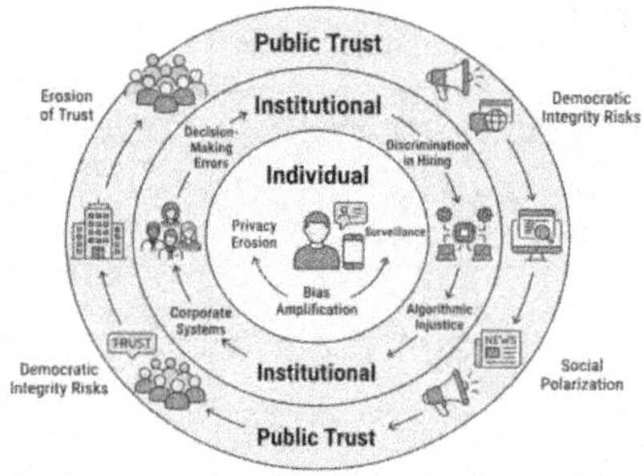

The societal impact of AI can be understood through risks affecting individuals, organizations, and overall public trust.

2.2 Ethical Imperatives

Ethical imperatives in AI governance are not abstract principles. They are organizational commitments that must be translated into specific program requirements, control structures, and documented decisions. The challenge for managers is that ethics in AI is contested terrain: different stakeholders hold different values, different regulatory frameworks emphasize different principles, and the technical implementation of ethical requirements is genuinely difficult. This does not exempt the manager from the obligation to engage with the substance. It means the engagement requires both clarity about values and rigor about implementation.

Four ethical imperatives are foundational to AI governance programs. The first is non-maleficence: the obligation not to cause harm through AI-driven decisions. This is the most concrete imperative and the most tractable. It translates into requirements for bias testing, harm impact assessment, and the maintenance of human override capability in high-stakes decisions. A system that causes demonstrable harm is failing an ethical requirement that is simultaneously a legal and organizational liability. Non-maleficence is not a passive obligation; it requires affirmative action to identify and address harm pathways before they are activated.

The second imperative is beneficence: the obligation to design AI systems that produce genuine benefit for the people they affect, not just operational efficiency for the deploying organization. Beneficence shapes the framing of system objectives. A healthcare AI optimized to reduce hospital length of stay is not automatically beneficent; it is optimized for a metric that may or may not align with patient welfare. Beneficence requires asking whether the optimization target represents benefit from the perspective of the affected population, not just the deploying organization. In practice, this requires incorporating affected population perspectives into system design, which may involve structured community engagement, patient advisory input, or employee representation in setting system objectives.

The third imperative is autonomy: the obligation to preserve meaningful human agency in decisions that significantly affect individual welfare. Autonomy is eroded when AI recommendations are presented with false authority, when the system produces an output that users treat as definitive because they do not understand its limitations or have not been trained to question it. It is protected by human-in-the-loop design, transparent communication about AI's role in a decision, and mechanisms that allow affected individuals to request human review. Autonomy requirements are increasingly codified: the EU AI Act, the White House AI Bill of Rights, and several state-level frameworks include explicit human oversight and opt-out provisions that operationalize this imperative.

The fourth imperative is justice: the obligation to ensure that AI systems do not systematically disadvantage individuals or communities based on characteristics that are ethically irrelevant to the decision being made. Justice is the ethical foundation of bias and fairness requirements. It demands that governance programs go beyond checking whether protected characteristics are explicitly included in a model's inputs, because proxy variables can reproduce protected-class effects, and assess whether outcomes are equitable across relevant population groups. Justice also requires that governance processes themselves be equitable: that the people most affected by AI systems have meaningful opportunities to raise concerns and that those concerns receive genuine consideration.

Translating these imperatives into governance requirements means that each AI system's approval documentation should address how each imperative was evaluated. Not at an abstract level, not merely stating that the system was designed with fairness in mind, but with specific analysis: what populations were represented in the training data, what fairness metrics were computed, what harm scenarios were tested, what human oversight mechanisms exist, and what the system's objective function optimizes for and on whose behalf. The approval documentation is the governance record of the organization's ethical reasoning, and it must be specific enough to be meaningful to a reviewer who was not present when the decisions were made.

Diagram 2.3: Four Ethical Imperatives Mapped to Governance Controls

Ethical Imperative	Governance Question	Required Control
Non-maleficence	Does the system pose risks of harm to stakeholders? *Determine Risk*	Implementation of risk assessment and mitigation protocols.
Beneficence	How are positive outcomes and stakeholder benefits maximized? *Define Benefits*	Clearly defined benefit analysis and outcome monitoring metrics.
Autonomy	Is individual autonomy and choice respected? *Design for Choice*	Process for informed consent and data privacy rights.
Justice	Is the distribution of benefits and burdens equitable? *Audit for Fairness*	Audits for bias detection and fairness in resource allocation.

Caption: The four actions-imperatives mapped to matte that could be devidebring ethical imperatives mapped to governance controls

2.3 Stakeholder Mapping

Stakeholder mapping is the process of identifying everyone who has a material interest in an AI system—those affected by its outputs, those who depend on its performance, those responsible for its governance, and those with the power to challenge or constrain its operation. Mapping stakeholders before deployment is not a public relations exercise. It is a risk management discipline: stakeholders who are not identified before a system goes live will surface afterward, often in adversarial contexts that could have been avoided with earlier engagement.

The primary stakeholder categories for most enterprise AI systems include: First, affected individuals: the people whose lives, opportunities, or conditions are directly shaped by the system's outputs. In a loan underwriting system, these are the applicants. In a healthcare AI, these are the patients. In a hiring tool, these are the job candidates. Their interests are represented in governance through bias testing, impact assessment, and the maintenance of human review and redress mechanisms. Affected individuals are often the stakeholder group with the least direct voice in the governance process and the most direct exposure to its consequences.

Second, operational users: employees who interact with the system's outputs in their work. A credit analyst who receives risk scores, a recruiter who works with ranked candidate lists, and a clinician who receives diagnostic recommendations are operational users. Their interests include accurate and reliable outputs, interpretable results, and clear guidance on when to follow or override the system's recommendation. Operational users who do not trust the system will work around it; those who trust it too completely will fail to catch its errors. Both failure modes represent governance risks that training and user interface design must address.

Third, organizational leadership: the executives and senior managers who are accountable for the business outcomes that AI systems affect and for the organization's compliance posture. Leadership stakeholders need regular reporting on AI system performance, risk status, and incident history. They need confidence that the governance program is functional, not merely documented. Governance programs that cannot produce meaningful reporting for leadership will fail to sustain organizational attention and resources when priorities compete with deployment velocity and cost management.

Fourth, regulators and legal authorities: the governmental bodies that have jurisdiction over the domains where AI is deployed. Their requirements define minimum governance standards, and their enforcement actions define the consequences of failure. Regulatory stakeholder analysis requires knowing which specific requirements apply to each AI system and what evidence regulators expect to see in an inquiry or audit. Regulatory expectations are evolving rapidly; the governance program must include a mechanism to track regulatory developments and update governance requirements accordingly.

Fifth, community and advocacy stakeholders: the civil society organizations, community groups, and public advocates who represent the interests of populations affected by AI systems. These stakeholders have growing capacity to surface AI failures through independent technical analysis, public advocacy, and regulatory complaints. A governance program that proactively engages with community perspectives before deployment through structured impact assessments, public comment processes, or community advisory input is better positioned than one that encounters community concerns for the first time in a news article or formal complaint.

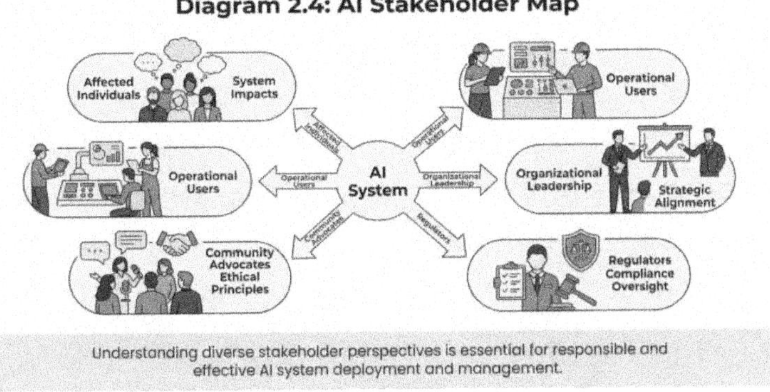

Understanding diverse stakeholder perspectives is essential for responsible and effective AI system deployment and management.

2.4 Social Risk Assessment

Social risk assessment is the structured analysis of how an AI system may produce adverse social outcomes that harm individuals or communities, beyond what is captured by purely technical performance metrics. It is conducted before deployment and reviewed at defined intervals during operation. It is distinct from and complementary to both technical risk assessments (which focus on model performance, reliability, and security) and legal risk assessments (which focus on regulatory compliance).

A social risk assessment asks three foundational questions. The first is: who bears the costs of this system's errors? All AI systems produce errors; the question is which populations are most likely to experience those errors, and whether those populations already face social or economic disadvantage. A credit model that misclassifies higher-income applicants creates moderate, recoverable harm, a declined loan that the applicant can contest or obtain elsewhere. A benefits eligibility system that misclassifies lower-income applicants creates severe, sometimes irreversible harm, loss of housing support, healthcare access, or nutritional assistance, with no practical recourse. The same error rate produces radically different social risk depending on who bears the cost.

The second question is: Does this system concentrate or distribute benefits? Systems that route resources, services, or opportunities create distributive effects. Optimization models that maximize aggregate efficiency can systematically concentrate benefits among already well-served populations while reducing services to those who need them most. The transit authority example at the opening of this chapter illustrates this precisely: route optimization maximized aggregate ridership efficiency while distributing service reductions onto the neighborhoods least able to absorb them. Social risk assessment requires explicitly mapping these distributive effects, not assuming that aggregate optimization aligns with equitable distribution.

The third question is: what power relationships does this system create or reinforce? AI systems deployed in asymmetric relationships between an employer and an employee, between a government agency and a benefit recipient, between a lender and a borrower have the potential to amplify the power of the organization deploying them. The automated nature of the decision may make it harder for individuals to understand, contest, or influence outcomes. Social risk assessment requires identifying these power dynamics and ensuring that governance controls explainability requirements, human review mechanisms, and formal redress processes provide a meaningful counterbalance.

The output of a social risk assessment is a documented analysis that names specific social risks, assigns each a severity rating and likelihood estimate, identifies the affected populations, describes the proposed controls, and names an owner responsible for monitoring. This document becomes part of the system's governance record and is reviewed when the system undergoes a significant change or when its operating context shifts materially. Social risks that were acceptable in one context may become unacceptable as community expectations, regulatory standards, or organizational commitments evolve.

Diagram 2.5: Social Risk Assessment Framework

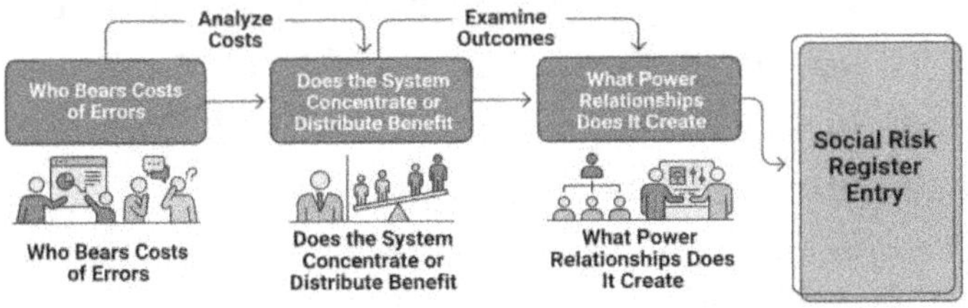

Systematic evaluation of societal impacts identifies risks for mitigation strategies.

2.5 Equity and Fairness

Equity and fairness in AI are among the most technically and organizationally complex governance requirements. They are also among the most consequential: a system that produces systematically inequitable outcomes is not merely an ethical problem; it is a legal liability under anti-discrimination law, a reputational risk, and an operational failure that substitutes statistical precision for justice in the decisions that matter most to the people it affects.

The technical complexity arises from the fact that fairness is not a single, well-defined property. Multiple mathematical definitions of fairness exist, and they are frequently in tension with each other. Demographic parity requires that positive outcomes occur at equal rates across protected groups. Equalized odds require that both true positive rates and false positive rates be equal across groups. Predictive parity requires that the model's predictions carry the same meaning across groups: a predicted score of 0.7 indicates the same probability of the outcome regardless of which group the individual belongs to. These definitions cannot all be satisfied simultaneously in most real-world settings. The governance requirement is not to satisfy all fairness metrics but to make a documented, accountable decision about which fairness standard applies to the specific decision context, and why, and to demonstrate that the chosen standard is met.

The organizational complexity arises from the fact that achieving equitable AI outcomes requires confronting historical inequity in data. Training data that reflects decades of discriminatory practices encodes those practices into model parameters. Correcting for that bias requires deliberate interventions, such as reweighting training samples, adjusting decision thresholds, and constraining model features that are technically nontrivial and may be organizationally contested. Some stakeholders will frame these interventions as reducing model accuracy. The governance response is to clarify that accuracy measured on biased historical outcomes is not the correct standard, and that a model can be highly accurate by statistical measures while being deeply unjust in its social effects.

For managers, the equity requirement means that bias and fairness analysis must be a mandatory checkpoint in the AI development and deployment process, not an optional enhancement. The analysis must be documented, must identify the specific fairness standard applied and the reason for that choice, must show test results disaggregated by relevant population groups, and must identify any residual disparity and either the rationale for accepting it or the plan for addressing it. A deployment approval that does not include a fairness analysis is incomplete governance.

Equity also requires ongoing monitoring after deployment. A model that was equitable at launch may become less equitable over time as the population it serves changes, as feature distributions shift, or as the social context in which it operates evolves. Fairness metrics should be included in the monitoring framework alongside accuracy and operational metrics. A degradation in fairness measures is as significant a performance concern as a degradation in predictive accuracy, and it should trigger the same review-and-response process.

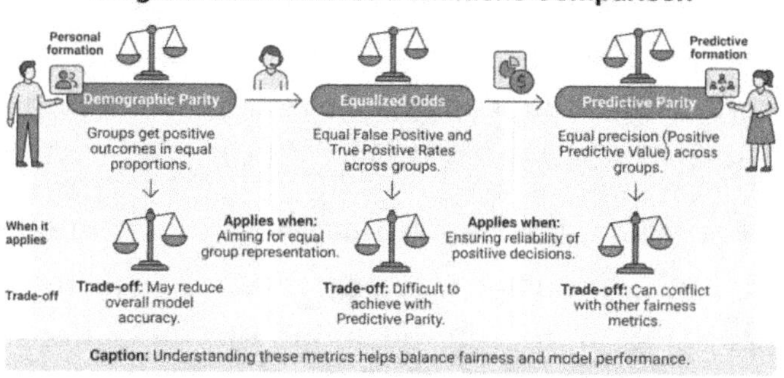

Diagram 2.6: Fairness Definitions Comparison

Caption: Understanding these metrics helps balance fairness and model performance.

2.6 Community Expectations

Community expectations represent the informal but powerful governance requirements that exist outside formal regulatory frameworks. They reflect what affected populations, advocacy organizations, civic institutions, and the general public believe is acceptable conduct when organizations use AI to make or influence consequential decisions. These expectations vary by context, culture, and the history of an organization's relationships with the communities it serves, but several patterns hold broadly across enterprise AI deployments.

Transparency is the most consistently held community expectation. People who are subject to AI-driven decisions want to know that a decision was made by or with the assistance of an AI system, what the basis for the decision was, and what recourse exists if they disagree with the outcome. This expectation has increasingly been codified in regulations, the GDPR's right to explanation, the EU AI Act's transparency requirements for high-risk systems, and U.S. federal agency guidelines on automated decision-making, all of which reflect community expectations formalized as legal requirements. But even where formal requirements do not exist, organizations that operate opaquely in communities with significant power relationships accumulate trust deficits that eventually manifest as regulatory pressure, organized resistance, or reputational events.

Accountability is the second pattern. Community stakeholders expect that when an AI system causes harm, a responsible person or institution can be identified, will acknowledge the harm, and will take corrective action. This is the governance requirement of named ownership made externally visible, not diffuse organizational responsibility, but specific, named accountability that makes meaningful response possible. An organization that responds to a community harm by diffusing responsibility across multiple teams and time periods is not meeting the accountability expectation; it is confirming the community's concern that no one is actually in charge.

Meaningful participation is a less universal but growing expectation, particularly in the public sector and heavily regulated contexts. It holds that communities significantly affected by AI systems should have an opportunity to provide input before those systems are deployed. This does not require elaborate public consultation processes for every AI tool. It does require that governance programs for high-impact systems include structured impact assessment that incorporates the perspectives of affected communities, and that feedback mechanisms exist for affected populations to raise concerns after deployment. Organizations that build participation mechanisms proactively find that they surface concerns earlier and generate more trust with the communities they serve.

The governance response to community expectations is not a communications strategy; it is a program design strategy. Transparency is achieved through explainability controls and disclosure policies built into the system. Accountability is achieved through named ownership and defined response processes. Meaningful participation is achieved through structured community engagement in the impact assessment process. Each of these is a governance artifact, not a public affairs activity.

2.7 Manager's Ethical Accountability

The concept of ethical accountability is often treated as a cultural or values-based aspiration that organizations should be ethical, and managers should behave responsibly. That framing understates the practical stakes. In the AI context, ethical accountability has become an operational and legal category. Managers who deploy AI systems that produce discriminatory outcomes, violate privacy, or cause community harm without adequate governance are exposed to regulatory sanction, civil liability, and organizational consequences. Ethical accountability is governance accountability made specific.

The manager's ethical accountability has three dimensions. The first is anticipatory: the obligation to identify potential social harms before deployment and to put in place controls that reduce those harms to acceptable levels. This is the function of stakeholder mapping, social risk assessment, and equity analysis conducted during the system design and approval process. A manager who proceeds with deployment without conducting those analyses is not taking a calculated risk; they are creating undocumented exposure that cannot be defended in a subsequent investigation or audit. The anticipatory obligation cannot be outsourced to the technical team or delegated to the compliance function. It is the manager's accountability because the manager owns the business process and bears responsibility for its outcomes.

The second dimension is ongoing: the obligation to monitor deployed systems for emerging social harms and to act when monitoring reveals adverse outcomes. A system that was equitable at deployment may become inequitable as the world changes, as the population it serves evolves, or as its operating parameters drift. Ongoing monitoring is not a luxury for well-resourced governance programs; it is the mechanism that converts a deployment decision made in good faith into a sustained commitment to responsible operation. A manager who approved deployment and then never checked whether the system continued to behave appropriately has not fulfilled the ongoing obligation.

The third dimension is corrective: the obligation to act decisively when social harm is identified, even when acting is organizationally costly. Corrective action may mean adjusting a model's decision thresholds to reduce disparity, restricting the system's application to contexts where it has been validated, suspending operation pending a thorough review, or communicating proactively with affected individuals. The manager who discovers social harm and takes no action, or who acknowledges a concern and responds with delay while exposure accumulates, bears accountability for the interval of continued harm. The governance record must show that harm identification triggered a substantive and timely response, not just documentation of the concern.

These three dimensions, anticipatory, ongoing, and corrective, define the full arc of ethical accountability in AI governance. A manager who performs them consistently across every AI system in scope is not merely complying with governance requirements. They are exercising the professional judgment that responsible AI management demands: applying analytical rigor to social questions, representing the interests of populations that are not in the room when deployment decisions are made, and accepting accountability for outcomes rather than merely for activities.

2.8 Manager's Checklist

The following checklist operationalizes the social governance requirements introduced in this chapter. It is organized as a pre-deployment assessment to be completed before any AI system that affects external individuals or communities is approved for production.

Societal impact scope: Document who is directly affected by this system's outputs, at what scale, and with what potential severity. Identify whether any affected population has pre-existing social or economic disadvantages that amplify the harm caused by system errors. This documentation should be reviewed by someone outside the deployment team before approval.

Ethical imperative review: For each of the four ethical imperatives, non-maleficence, beneficence, autonomy, and justice, document the specific question it poses for this system and the specific control that addresses it. Absence of an answer is a governance gap that must be resolved before deployment approval. A table format is appropriate: one row per imperative, one column for the governance question, one column for the control.

Stakeholder map: Complete and document a stakeholder map for this system. Identify at minimum: affected individuals, operational users, organizational leadership stakeholders, applicable regulators, and relevant community or advocacy stakeholders. For each group, document the governance mechanism through which their interests are represented in the deployment decision.

Social risk assessment: Conduct and document a social risk assessment addressing cost distribution, benefit distribution, and power relationship effects. Assign a severity rating to identified social risks and document the controls in place. Open risks that are without adequate controls must be resolved or accepted through the formal risk acceptance process before deployment.

Fairness analysis: Select and document the applicable fairness standard for this decision context and explain the basis for the selection. Show test results disaggregated by relevant population groups. If residual disparity exists, document the rationale for accepting it or the remediation plan, including a timeline and owner.

Transparency commitment: Define what information will be provided to affected individuals about the AI system's role in decisions affecting them, and through what mechanism. Ensure that the information is actionable; it should enable individuals to understand the basis for a decision and to seek human review if desired.

Monitoring schedule: Define the social monitoring indicators to track for this system outcome disparities, complaint rates, escalation patterns, and set a review schedule appropriate to the system's risk level. High-impact systems require a quarterly review at a minimum. Monitoring must be conducted on schedule, not merely scheduled.

2.9 What This Chapter Establishes

AI systems are social interventions, not merely technical tools. The manager who treats AI governance as a purely technical or compliance exercise will fail to manage its most consequential risks. These risks emerge from the way AI systems reshape decisions, distribute benefits and burdens, and affect the communities they touch. Those risks are not hypothetical: they appear in regulatory enforcement actions, civil litigation, community organizing, and the kind of public controversy that consumes organizational resources for years after the fact.

The governance framework this chapter introduces societal footprint analysis, operationalized ethical imperatives, stakeholder mapping, social risk assessment, fairness analysis, transparency and accountability mechanisms, and the anticipatory-ongoing-corrective accountability structure provides the analytical foundation for governing those risks. It connects the abstract ethical principles that appear in regulatory guidance and organizational values statements to the concrete program requirements and management behaviors that make governance functional.

The chapters that follow build on this foundation to address governance structures, legal frameworks, and operational controls. But those structures and controls only function as intended when managers recognize what they are governing: systems that act in the world, on people, with consequences that extend beyond the organization's operational boundaries. That recognition that AI governance is ultimately about the relationship between organizations and the communities whose lives AI systems affect is the perspective this chapter establishes.

3 AI Governance

A financial services firm had grown its AI footprint incrementally over five years. Individual business lines had procured or built their own models for credit risk, fraud detection, customer segmentation, marketing optimization, and document processing, each under the oversight of the line's head of technology or analytics. By the time the enterprise risk committee asked for a consolidated view of the firm's AI exposure, no single person could produce it. The existing inventory was incomplete, the documentation created was inconsistent, and accountability for any given system was either contested or undefined. When the first regulatory examination specifically targeting AI practices arrived, the firm spent nine months and incurred significant external consulting fees to construct a governance program retroactively. The cost was not just financial; it was a period of genuine exposure during which systems operated without the oversight structure that should have been in place from the start.

That scenario captures the central challenge of AI governance: it must be built deliberately, not assembled after the fact. The governance structures that make AI accountable, with clear roles, enforced policies, documented decisions, and defined escalation paths, do not emerge spontaneously from technically excellent teams. They must be designed, resourced, and maintained as organizational infrastructure. This chapter provides that design.

The sections that follow address the three structural choices every organization must make: what governance model to use, how to assign roles and accountabilities, and what committees and escalation mechanisms to establish. They then address the operational requirements that determine whether governance is enforceable: decision rights frameworks, auditability, and integration with existing IT and compliance functions. The goal is not merely a governance chart on an intranet page but a governance program that functions under the operational pressures of a real enterprise.

Diagram 3.1: AI Governance Architecture Overview

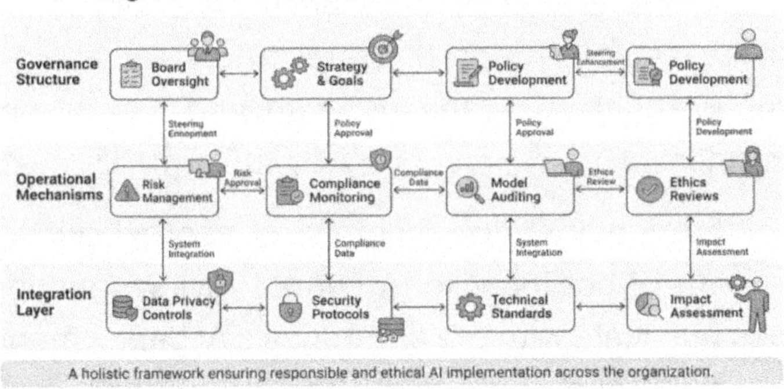

3.1 Governance Models

The governance model determines how authority over AI systems is organized across the enterprise. Three models dominate practice: centralized, federated, and hybrid. The choice among them is not primarily a technical decision; it reflects the organization's size, regulatory environment, the maturity of its AI program, and the degree of consistency required in the management of AI risks.

A centralized governance model places authority for AI policy, risk assessment, and deployment approval in a single organizational unit, typically reporting to a Chief AI Officer, Chief Data Officer, or Chief Risk Officer. All AI systems across the enterprise require review and approval from this central function before deployment. Policy is set centrally and applies uniformly. Monitoring and audit are conducted centrally or under central supervision. The principal advantage of centralization is consistency: the same standards apply everywhere, accountability is clear, and the enterprise has a single, coherent view of its AI risk exposure. Centralization also makes regulatory reporting straightforward. There is one function that owns the program and can respond to inquiries.

The principal disadvantage of centralization is the creation of bottlenecks. A central review function that does not scale with the enterprise's AI deployment pace becomes a chokepoint that either delays legitimate work or is circumvented through informal channels. Organizations that adopt centralized governance must invest in the central function's capacity to match deployment demand. A governance function that is perpetually behind and processes reviews on multi-month timelines will lose organizational credibility and become a compliance-theater exercise rather than a functioning oversight mechanism.

A federated governance model distributes authority to business lines or functional units, each of which operates its own AI governance function under enterprise-level policy guidance. Each unit is responsible for its own risk assessment, documentation, and monitoring. The enterprise sets minimum standards and may operate a center of excellence that provides guidance and tools, but approval authority sits with the business line. The principal advantage of federation is speed and contextual relevance: the people who review AI systems in a specific domain have the domain knowledge to assess risks meaningfully. A credit risk reviewer who understands underwriting can conduct a more meaningful model review than a centralized generalist.

The principal disadvantage of federation is inconsistency. Without strong enterprise-level policy enforcement, federated governance programs tend to produce uneven standards, with some units maintaining rigorous programs and others applying minimal oversight. Federated governance also makes it difficult to aggregate enterprise-level risk. When each business line manages its own AI risk, producing a consolidated view of the enterprise's exposure requires significant coordination that may not occur unless explicitly mandated.

A hybrid governance model combines central standards with distributed execution. Enterprise policy and risk taxonomy are set centrally. Minimum documentation, mandatory review checkpoints, and escalation thresholds are centrally defined and enforced. But day-to-day governance activities, risk assessment, monitoring, and documentation are carried out by embedded governance leads or AI risk managers within each business line or functional unit. This model requires more coordination infrastructure than either pure alternative, but it is the most scalable approach for enterprises that need both consistency and deployment velocity.

The choice of model should be made explicitly and documented. Organizations that have not made a deliberate choice typically end up with informal centralization, where a central team reviews high-profile systems. At the same time, business lines govern everything else without visibility, which combines the disadvantages of both approaches. A documented governance model establishes clear expectations about who reviews what, who approves deployment, and where accountability sits when things go wrong.

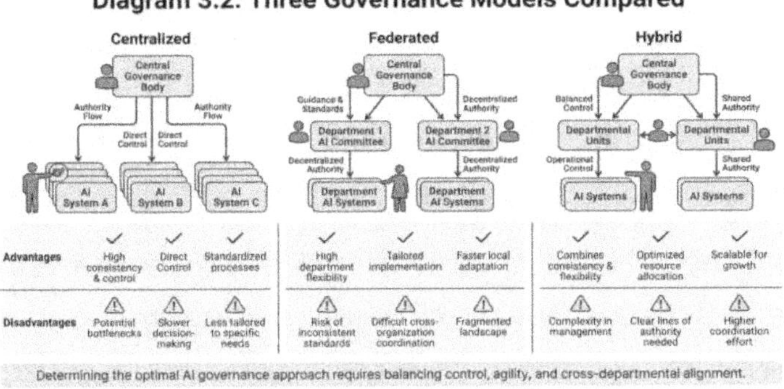

Diagram 3.2: Three Governance Models Compared

3.2 Roles and Accountabilities

A governance model is a structural choice; roles and accountabilities are how that structure becomes operational. Without named owners and explicit responsibilities, governance frameworks remain notional; the right activities are described, but no specific person is obligated to perform them. Every AI system in the enterprise should have a clearly defined set of roles covering its approval, operation, and oversight.

The AI system owner is the individual responsible for the system's performance, compliance, and governance. This is typically a senior manager in the business unit or functional area that uses the system, not the technical team that built or configured it. The system owner approves deployments, reviews monitoring reports, authorizes significant changes, and serves as the first point of escalation for governance concerns. The owner role must be assigned to a named person, not a team, not a title, not a function, because distributed accountability evaporates in a crisis. When a system fails, the first question from regulatory and senior leadership is who owns this, and that question must have an unambiguous answer.

The AI developer or model owner is the technical person or team responsible for building, configuring, and maintaining the AI system. Their governance accountabilities include producing required documentation (model cards, data provenance records, validation test results), implementing technical controls specified by the governance review, and flagging performance issues or drift to the system owner. The developer role often sits in an engineering or data science function; governance must ensure that documentation and control requirements are embedded in the development workflow rather than treated as end-of-process compliance activities that are rushed through after technical work is complete.

The AI risk and compliance function provides independent oversight. It maintains the enterprise risk taxonomy for AI, conducts or coordinates periodic reviews of AI systems in production, tracks regulatory developments that affect governance requirements, and reports to senior leadership and the board on the enterprise's aggregate AI risk exposure. In small organizations, this function may be handled by a single person embedded within the existing risk or compliance team. In large enterprises, it may be a dedicated AI governance office. What matters is that the function is genuinely independent of the business lines it oversees; a compliance function that reports to the business head is supposed to review, is structurally compromised, and will produce governance shaped by the interests it should be checking.

The data governance function intersects with AI governance at the point where training, inference, and model outputs are managed. Data governance is responsible for data classification, access controls, lineage documentation, and privacy compliance. AI governance depends on effective data governance: a model trained on unclassified, poorly documented data poses privacy and lineage risks that AI governance cannot resolve on its own. The two functions must be coordinated, with clear handoffs and shared accountability for data-related AI risks.

Executive sponsorship is the role that ensures AI governance has the organizational authority and resources it needs. The executive sponsor, typically a member of the senior leadership team, advocates for governance investment, resolves cross-functional disputes about governance requirements, and signals to the organization that governance is a genuine priority rather than an optional overhead. Governance programs without executive sponsorship tend to be under-resourced and marginalized in priority-setting conversations. The sponsorship role is most critical during the program's first two years, when governance requirements are new, and the organizational reflex to treat them as secondary to deployment velocity is strongest.

Diagram 3.3: AI Governance Role Map

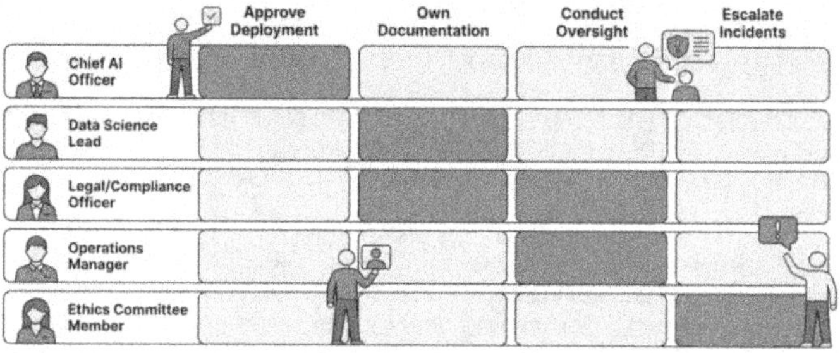

This role map clearly outlines key AI governance responsibilities to ensure accountability and compliant operations across the organization.

3.3 Governance Committees

Governance committees are the collective decision-making mechanisms required by AI governance. Not every AI decision warrants committee review that would be operationally paralytic, but the decisions that carry significant risk, require cross-functional coordination, or set policy precedent should be made through a defined committee process rather than by individual judgment alone. Committees create accountability, enable cross-functional input, and produce documented decisions that support audit and regulatory review.

An AI risk committee is the central governance body for AI decisions at the enterprise scope. Its membership typically includes the Chief Risk Officer or equivalent, the head of the AI governance function, representatives from legal and compliance, and senior representatives from major business lines that deploy AI. Its mandate covers approving governance policies, reviewing significant AI deployments or changes, adjudicating governance escalations, and overseeing the enterprise's aggregate AI risk posture. The AI risk committee should meet on a regular cadence, monthly or bimonthly, for most enterprises. It should maintain a formal charter that specifies its authority, composition, quorum requirements, and decision-making procedures.

A model review board is the operational-level committee that reviews individual AI systems before deployment and periodically during operation. It provides the detailed technical and risk review that the AI risk committee is not positioned to conduct at volume. The model review board's membership typically includes representatives from data science or engineering, risk management, legal, and the relevant business line. Its output includes a deployment approval or conditional approval with required remediations, a record of the evidence reviewed, and a monitoring plan specifying the indicators and schedule for ongoing review. The existence of a model review board with documented review records is one of the most important evidence artifacts in a regulatory examination.

An ethics advisory committee or equivalent body provides a structured forum for reviewing AI applications that raise significant social, ethical, or equity concerns. Not every organization needs a separate ethics committee; some embed this review in the model review board, but organizations that deploy AI in high-stakes domains (healthcare, criminal justice, employment, financial services, public benefits) benefit from a dedicated mechanism that applies the social governance framework from Chapter 2 to specific deployment decisions. An ethics advisory committee with external representation from civil society, affected communities, or academic expertise provides a degree of independence that internal review cannot fully replicate, and its existence demonstrates to regulators and the public that the organization takes equity and social impact seriously as governance requirements rather than values statements.

Committee effectiveness depends on two operational requirements. The first is documented authority: each committee must have a written charter that specifies what decisions it makes, what information it reviews, and what constitutes a valid decision. Without documented authority, committees become advisory rather than governing, and their decisions are not binding. The second is accountability for follow-through: decisions made in committee must be tracked to completion. A committee that approves deployment subject to remediation conditions but never verifies that conditions were met is generating governance theater, not governance. Condition tracking should be a standing agenda item at each meeting, with overdue conditions escalated to the committee chair for resolution.

Diagram 3.4: Governance Committee Structure

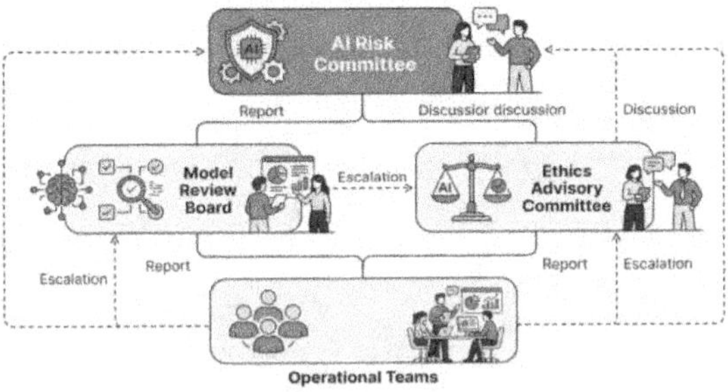

'This framework outlines the hierarchical structure for responsible AI oversight and decision-making.'

3.4 Escalation Paths

An escalation path is the defined route by which a governance concern, a detected anomaly, a compliance question, a potential harm, or an unresolved dispute about AI system behavior moves from the point of initial identification to the appropriate decision authority. Escalation paths are among the most underinvested elements of AI governance programs. Organizations invest in policies, committees, and documentation frameworks, then leave escalation to informal judgment. The person who spots a problem figures out who to tell. That approach fails precisely when it matters most: in a crisis, under time pressure, with organizational incentives pulling against disclosure.

A functional escalation path has four elements. The first is trigger definition: a clear description of what conditions require escalation. Triggers should include detected performance degradation below defined thresholds, evidence of discriminatory outcomes, data handling violations, unresolved disputes between the system owner and the risk function, and any incident that has or may have caused harm to an individual or community. Triggers should be specific enough to be actionable. A trigger defined as "significant concerns" provides no operational guidance. Every person in the governance chain should be able to read the trigger definition and, without interpretation, determine whether a given situation requires escalation.

The second element is the escalation chain: the sequence of roles to which a concern is escalated and the time frame for each level's response. A typical chain runs from the person who identifies the concern to the system owner, from the system owner to the AI risk function, from the AI risk function to the committee with authority to order a response, and from the committee to executive leadership if the concern rises to enterprise significance. Time frames must be defined an escalation path that moves at organizational speed is inadequate for harms that compound over time. High-severity incidents should reach executive awareness within hours, not days.

The third element is documentation: a requirement that each escalation is recorded, including what was escalated, when, by whom, to whom, what decision was made, and what action was taken. This record serves two functions. Internally, it supports accountability by ensuring that people who escalate know their escalation is documented, and decision-makers know their responses are on record. Externally, it provides the evidence trail that demonstrates governance was exercised when a concern arose. In a regulatory examination, the escalation record is often the most direct evidence of whether a governance program was functional or aspirational.

The fourth element is protection for good-faith escalators: an explicit organizational commitment that employees who raise AI governance concerns in good faith will not face retaliation. AI governance depends on the people closest to the systems engineers, analysts, and operational users being willing to surface concerns. If organizational culture or management behavior signals that raising concerns creates personal risk, the escalation path will be unused. The governance program must actively signal that surfacing concerns is expected and valued, not treated as disloyalty or disruption. Some organizations formalize this through a governance reporting mechanism with a protected channel, similar to a compliance hotline.

Diagram 3.5: AI Escalation Path

A formal framework for reporting and resolving risks in Artificial Intelligence systems across enterprise levels.

3.5 Decision Rights Frameworks

Decision rights frameworks define who has authority to make which AI governance decisions. They address a persistent governance failure mode: ambiguity about authority leads to either conflict, multiple parties believing they have authority and disagreeing, or a vacuum in which no party is willing to decide because accountability is unclear. A decision rights framework resolves that ambiguity in writing, before decisions need to be made, so that when pressure is applied, the response path is defined rather than improvised.

The RACI model Responsible, Accountable, Consulted, Informed) is the most widely used tool for documenting decision rights. In the AI governance context, it is applied to a defined set of decision categories: system deployment approval, significant model change approval, policy development, risk acceptance, incident response authorization, vendor selection and contracting, data access authorization, and system decommission. For each decision category, the RACI matrix specifies which role is Responsible (does the work of preparing the decision), Accountable (owns the outcome and cannot delegate), Consulted (must provide input before the decision is finalized), and Informed (receives the decision outcome but does not influence it).

A common governance failure occurs when accountability is assigned at a level that is too senior to be practically exercised. An executive who is nominally accountable for all AI deployments cannot, in practice, review each one with meaningful attention. Decision rights frameworks must assign accountability at the appropriate operational level based on the decision's risk and complexity, with escalation to senior levels reserved for decisions that meet defined thresholds by risk rating, cost, regulatory exposure, or scope of impact. The framework must be operationally realistic to be operationally functional.

Decision rights frameworks also address boundary conditions: decisions that cross organizational boundaries and require coordination across functions or business lines. AI decisions that involve data owned by one function and deployed in another, that affect workflows managed by multiple teams, or that carry legal obligations reviewed by counsel. At the same time, technical responsibility lies with engineering; these boundary-crossing decisions are particularly prone to confusion over accountability. The decision rights framework should identify each category of cross-boundary decision and specify the coordination mechanism, the single accountable party, and the escalation path if coordination fails to produce a timely decision.

The practical artifact is a RACI matrix for AI governance decisions, maintained as a living document and reviewed at least annually or when organizational structures change. The matrix should be accessible to everyone in the governance structure, not locked in an executive's documentation system. It should be referenced explicitly in governance training and onboarding for new AI projects. A decision rights framework that no one can find is not functioning as a governance instrument.

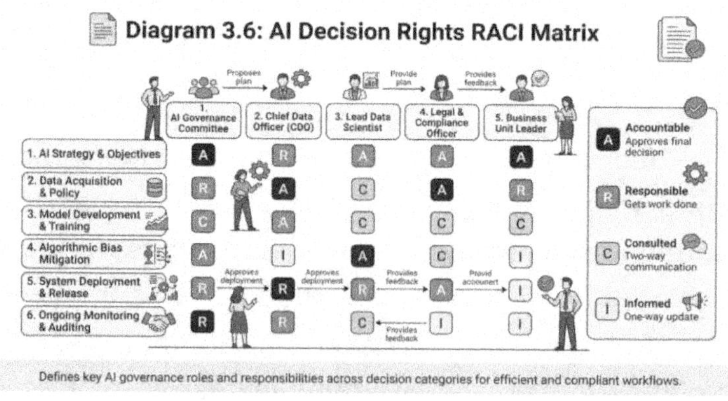

Defines key AI governance roles and responsibilities across decision categories for efficient and compliant workflows.

3.6 Enforceability and Auditability

A governance framework that cannot be enforced or audited is governance in name only. Enforceability requires that governance requirements have teeth, that non-compliance produces defined consequences, and that compliance is verified, not assumed. Auditability requires that governance activities produce evidence, documented decisions, approvals, reviews, and monitoring results that can be assembled in response to an audit or regulatory inquiry without extraordinary effort.

Enforceability begins with policy design. Governance policies must specify not just what is required but what happens when requirements are not met. Policies that describe ideal practices without specifying compliance obligations or consequences are aspirational documents, not governance instruments. Minimum enforceability elements include a clear statement of who the policy applies to, what the required behaviors are, what documentation must be produced, who verifies compliance, and what escalation or consequence applies when requirements are not met. Vague consequence language, such as 'appropriate disciplinary action,' is better than nothing, but specific consequence language is preferable: the deployment will be suspended pending a remediation review; the vendor contract will be subject to additional controls.

Enforceability also requires integration with existing enforcement mechanisms. Most organizations have compliance programs, audit functions, and HR processes that address policy violations. AI governance policies should be incorporated into those mechanisms rather than creating parallel enforcement structures. An AI governance policy violation that is handled through the same compliance process as any other policy violation carries organizational credibility that an AI-specific enforcement process may not initially have. Integration also reduces the administrative overhead of governance enforcement, which is particularly important in environments where governance functions are under-resourced.

Auditability is achieved through systematic documentation practices. For each AI system, the governance record should contain: the initial risk assessment and deployment approval, the model card or equivalent technical documentation, the data provenance record, the fairness analysis, the approval decision with the name of the approving authority and the date, a log of significant changes and their approvals, monitoring reports for each review period, and a record of any incidents, escalations, and their resolutions. This documentation does not need to be maintained in a bespoke system; it can be organized in a standard document management platform, but it must be organized so that a reviewer can reconstruct the governance history of any system without manual effort.

Auditability standards should be calibrated to the system's risk tier. A low-risk internal tool requires less documentation than a high-risk customer-facing model operating in a regulated domain. But minimum standards should apply to all systems: even low-risk tools should have a deployment record and an assigned owner. The question a governance program must be able to answer for any AI system at any time is: who approved this, when, based on what review, and who is responsible for it today? An inability to answer any of those questions is an auditability gap that must be remediated.

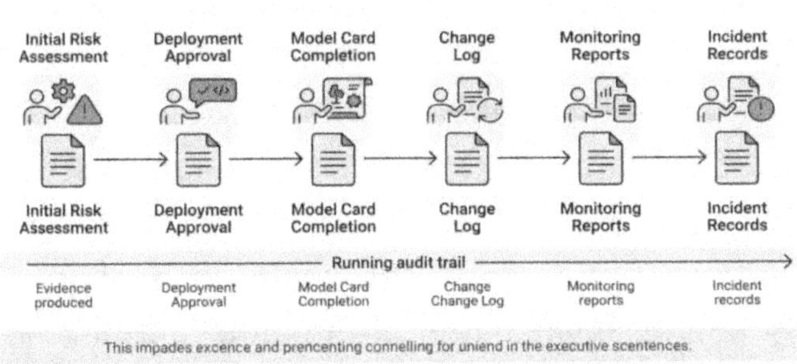

Diagram 3.7: Governance Documentation Lifecycle

3.7 Integration with IT and Compliance

AI governance does not need to be built from scratch. In most organizations, existing IT governance and compliance frameworks already provide infrastructure risk assessment processes, change control mechanisms, vendor management practices, and audit functions that AI governance can and should leverage. The failure to integrate AI governance with these existing functions is one of the most common and costly governance mistakes: it creates duplicative overhead, inconsistent standards, and resentment among technical and operational teams who are asked to satisfy multiple overlapping review requirements for the same system.

Integration with IT governance means incorporating AI systems into the existing IT risk management framework. AI systems are software systems and infrastructure components. They should appear in the IT asset inventory. They should be subject to change control. A significant model update should go through the same change review process as any other significant system change, potentially with additional AI-specific checkpoints to revalidate and reassess fairness. They should be subject to vendor management oversight when they are procured externally. They should be included in business continuity planning. The AI governance layer adds AI-specific requirements — fairness analysis, model documentation, bias testing — on top of this existing foundation, rather than replacing it.

Integration with compliance means ensuring that AI governance requirements are mapped to the compliance obligations that generate them. Privacy compliance, anti-discrimination law, sectoral regulations, and emerging AI statutes all impose requirements that governance controls must satisfy. The compliance function's regulatory monitoring capability should extend to AI-specific developments, and the governance program's documentation practices should be designed to produce the specific evidence that regulators in each applicable domain have indicated they expect. When a new AI regulation takes effect, the integration of AI governance and compliance should enable the organization to identify, within a defined time frame, which governance changes are required and who is responsible for implementing them.

Integration also requires coordination at the process level. When a new AI system is proposed, the review process should incorporate both IT governance checkpoints and AI governance review without requiring the development team to navigate two entirely separate processes. A unified review process that incorporates AI-specific steps within the existing IT project approval workflow reduces friction and increases adoption. Governance perceived as an external imposition on development teams is systematically worked around; governance embedded in the workflow teams already follow is more likely to be executed consistently and with genuine engagement rather than grudging compliance.

The practical mechanism for achieving this integration is joint ownership of the AI governance process between the AI governance function and the IT governance function. That does not require an organizational merger; these functions can remain distinct, but it does require that they operate from a shared process map, a common risk taxonomy, and coordinated communication to development teams and business lines. Integration should be documented in the governance program charter, and the integration design should be reviewed annually to ensure that it remains effective as both the AI program and the IT governance framework evolve. Integration that was designed for a governance program managing a dozen AI systems may not function for a program managing hundreds.

3.8 Manager's Checklist

The following checklist operationalizes the governance structural requirements introduced in this chapter. It applies to any manager responsible for building or improving an AI governance program or for a function that operates AI systems.

Governance model selection: Document the governance model in effect for your scope: centralized, federated, or hybrid. If the model has not been formally chosen, take that decision to the appropriate authority. Identify who is responsible for governance policy, who has deployment approval authority, and where the escalation path leads when those functions disagree. The absence of a documented model is itself a governance gap.

Role assignment: Ensure that every AI system in your scope has a named system owner, a named developer or model owner, and a clear path to AI risk and compliance oversight. Role assignments should be documented in the governance record for each system, not just in organization charts. Confirm that each named role-holder understands their accountabilities and has the authority and resources to fulfill them.

Committee charter: If you are responsible for a governance committee, confirm that it has a written charter specifying its authority, composition, quorum, and decision procedures. If a committee exists without a charter, drafting one is the first governance action — an unchartered committee has no enforceable authority and no mechanism for producing binding decisions.

Escalation path documentation: Ensure escalation paths are documented, communicated to all system owners and operational users, and include defined triggers and timeframes. Test the path at least annually by asking the people responsible for AI systems whether they know how to escalate a concern and to whom to escalate it. If they cannot answer, the path is not functioning.

Decision rights clarity: Produce or obtain a RACI matrix for AI governance decisions within your scope. Identify any decision categories where accountability is ambiguous or contested, and resolve them through the appropriate governance channel before the ambiguity leads to a crisis.

Audit readiness: For each AI system in your scope, confirm that the governance documentation required for a regulatory review exists, is current, and is organized so that it can be assembled without unreasonable effort. Establish a documentation completeness target, for example, 90% of systems meeting or exceeding minimum documentation standards, and track progress against it.

IT and compliance integration: Confirm that AI governance requirements are reflected in existing IT change control, vendor management, and compliance processes. If AI systems are reviewed through a separate process that is not connected to IT governance or compliance, plan and execute integration within a defined time frame.

3.9 What This Chapter Establishes

Governance structures are the architecture of accountability. The governance model determines where authority is organized; roles and accountabilities determine who exercises it; committees and escalation paths determine how decisions are made and concerns are surfaced; decision rights frameworks resolve authority disputes before they become crises; enforceability and auditability requirements convert governance intentions into verifiable practice; and integration with IT and compliance ensures that AI governance is sustained organizational infrastructure rather than a parallel program that competes for attention and resources.

A governance program that combines all of these elements, deliberately designed, precisely documented, and integrated into the organization's existing management infrastructure, is the foundation for all the operational governance requirements addressed in subsequent chapters. Without this foundation, policies exist without enforcement, risk assessments exist without authority to act on them, and documentation exists without the organizational commitment to maintain it.

The financial services firm spent nine months in the opening scenario and substantial resources building what could have been built in the first year retroactively. The cost of that delay was not just financial; it was a prolonged period of regulatory exposure, organizational disruption, and erosion of the trust that governance exists to protect. With the structures described in this chapter in place, the firm could have responded to the regulatory examination by producing the current governance program documentation. That is the practical outcome this chapter enables: governance that is ready when it is needed, because it was built when it was not yet required.

4 The Legal Framework

4.1 Opening Scenario

A mid-sized regional bank had been running an AI-powered loan underwriting system for fourteen months when a state attorney general's office sent an inquiry letter. The letter cited the Equal Credit Opportunity Act and asked the bank to produce documentation demonstrating that its model's outputs did not produce disparate impact on protected classes. The compliance team had no model card, no bias testing log, and no audit trail showing when the model had last been evaluated. The legal team had never been briefed on the system. Three attorneys were engaged at emergency rates. The system was taken offline within 72 hours while the organization scrambled to reconstruct documentation it had never created in the first place. The total remediation cost exceeded the projected three-year savings from automation. None of this was inevitable. The failure was not technical; the model itself was defensible. The failure was managerial: no one had translated the organization's legal obligations into operational requirements before deployment.

This scenario repeats itself with predictable variation across sectors. A healthcare system discovers that its patient-prioritization algorithm has been producing racially skewed triage recommendations for two years without triggering any documented review. A large employer learns, through a discrimination complaint rather than internal monitoring, that its resume-screening tool has been filtering out candidates from historically Black colleges and universities at rates that diverge sharply from those of applicants from other institutions. In each case, the root problem is the same: a governance gap between the technical deployment of an AI system and the legal obligations that governed it from the moment it touched real people's lives. Closing that gap is the core purpose of understanding the legal framework.

4.2 Why the Legal Framework Matters

AI systems operate inside an existing body of law. That law does not care whether a human underwriter or an algorithm produced a discriminatory outcome; the obligation is the same. What changes with AI is the speed, scale, and opacity of decisions. An algorithm making 10,000 lending decisions per day creates legal exposure 10,000 times per day. A manager who does not understand the legal landscape cannot assess that exposure, assign the right controls, or make informed deployment decisions. Legal compliance is not a problem for the legal department. It is a management problem that legal counsel helps solve.

The legal framework for AI is not a single statute. It is a layered structure of federal laws, sector-specific regulations, state privacy statutes, emerging AI-specific legislation, and international frameworks that apply simultaneously depending on where a system operates, what data it processes, and what decisions it influences. A manager who waits for a compliance audit to understand this landscape is already behind. The purpose of this chapter is to equip managers with a working map of that landscape specific enough to trigger the right questions and document the right artifacts, without requiring a law degree.

Every section that follows translates legal concepts into operational requirements. The goal is not to make managers their own lawyers. The goal is to make managers useful clients of legal counsel: people who arrive with the right artifacts, ask the right questions, and understand what a regulator will look for in an inquiry. That capability is built through deliberate preparation, not through reaction to enforcement. The manager who understands the legal landscape before deployment spends time on governance. The manager who encounters it for the first time in a regulatory letter spends money on crisis response.

4.3 The Regulatory Landscape

The regulatory landscape governing AI in the United States is, as of this writing, primarily sectoral rather than comprehensive. There is no single federal AI law equivalent to Europe's AI Act. Instead, existing laws, many written decades before machine learning, are being applied to AI by regulators who interpret their mandates broadly. The Federal Trade Commission has invoked its authority under the unfair and deceptive practices provision to challenge algorithmic systems that produce discriminatory outcomes. The Consumer Financial Protection Bureau has issued guidance applying the Equal Credit Opportunity Act to credit-scoring models. The Equal Employment Opportunity Commission has signaled that automated hiring tools fall within the scope of Title VII analysis. Healthcare AI systems fall under the Health Insurance Portability and Accountability Act, the Food and Drug Administration's Software as a Medical Device pathway, and the Office for Civil Rights' anti-discrimination guidance.

This sectoral patchwork means that risk exposure depends heavily on industry vertical and the nature of the decision being automated. A human resources AI tool used by a healthcare employer with federal contracts may simultaneously need to satisfy EEOC anti-discrimination standards, OFCCP contractor obligations, HIPAA (if any health data touches the hiring process), and applicable state AI employment laws. The manager's first task is to identify which legal frameworks apply to each AI system, not as an academic exercise, but as a prerequisite for control selection and documentation planning.

Diagram 4.1: Regulatory Landscape Map

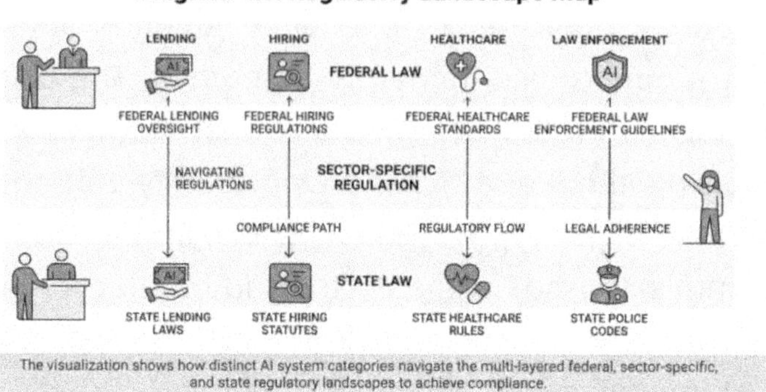

The visualization shows how distinct AI system categories navigate the multi-layered federal, sector-specific, and state regulatory landscapes to achieve compliance.

International exposure adds further complexity. Any organization processing personal data of European Union residents must contend with the General Data Protection Regulation regardless of where the organization is headquartered. The EU AI Act, which entered force in stages beginning in 2024, establishes explicit high-risk categories for AI applications, credit scoring, employment, critical infrastructure management, and biometric identification, each carrying specific transparency and human oversight requirements. Organizations operating globally cannot design AI governance programs based solely on the requirements of a single jurisdiction. The most restrictive applicable requirement typically sets the floor for the entire program.

The regulatory trend line is unambiguous regardless of current gaps. Every major regulatory body that has engaged with AI, the FTC, CFPB, EEOC, FDA, HHS Office for Civil Rights, SEC, and their counterparts in the EU, UK, Canada, and Australia, has moved toward more explicit requirements for transparency, documentation, human oversight, and fairness testing. Organizations that build governance programs to meet these emerging standards proactively will spend less on compliance over time than organizations that retrofit controls after mandates arrive. The regulatory landscape is not stable; it is directional. Managing in that direction is the smarter operational choice.

4.4 Privacy Laws

Privacy law is the most immediately operational legal constraint on AI systems. Every model that trains on, scores, or generates outputs about individuals processes personal data in a legally regulated way. The core frameworks in the United States are federal-sector laws, HIPAA for health data, the Gramm-Leach-Bliley Act for financial data, and the Family Educational Rights and Privacy Act for educational records, supplemented by a growing body of state privacy legislation. California's Consumer Privacy Act, as amended by the Consumer Privacy Rights Act, establishes the right to opt out of automated decision-making and requires businesses to conduct risk assessments for processing activities that pose significant privacy risks. Virginia, Colorado, Connecticut, Texas, and more than a dozen other states have enacted similar frameworks with varying specifics.

For managers, the operational implications of privacy law concentrate on three areas: lawful basis for processing, data subject rights, and data minimization. Lawful basis means the organization must be able to identify the legal ground under which it collects and uses personal data in AI systems, consent, contract, legitimate interest, or legal obligation, depending on the applicable framework. Data subject rights mean individuals may have the right to access data the organization holds about them, correct errors, request deletion, or opt out of automated processing that produces significant decisions about them. Data minimization means the model should be trained only on data that is necessary for its purpose, and data that is no longer needed should be purged on a defined schedule.

Diagram 4.2: Privacy Law Compliance Checklist Flow

Systematically verifying key requirements ensures data privacy law compliance.

The GDPR's automated decision-making provisions in Article 22 warrant specific attention from AI managers. When a processing activity produces decisions based solely on automated means that produce legal or similarly significant effects on individuals, the data subject has the right to human review, to contest the decision, and to an explanation of the logic involved. This is not a hypothetical provision; it has been enforced against credit, insurance, and hiring systems. Designing for Article 22 compliance means ensuring that automated outputs in high-stakes contexts are routed to human reviewers before they become final, and that the model's logic can be explained in plain terms to an affected person. These are architectural decisions that must be made before deployment, not after enforcement.

Privacy by design is the principle embedded in GDPR, adopted by reference in many state frameworks, and reflected in FTC guidance that privacy protections should be built into systems from the outset rather than appended after the fact. For AI systems, privacy by design has three primary implications. First, training datasets should be reviewed for data minimization before model development begins: does the model actually need all available data elements, or can it achieve equivalent performance on a reduced feature set that poses lower privacy risk? Second, the model's inference outputs should be designed to avoid exposing personal data that is not necessary for the output's purpose. Third, access controls on both training data and model outputs should be designed to the minimum necessary access level for each role that interacts with the system.

4.5 Sectoral Regulations

Sectoral regulations govern the industries where AI deployment carries the greatest consequences. Healthcare AI must navigate a particularly dense regulatory environment. The Food and Drug Administration's Software as a Medical Device guidance establishes a risk-based framework: AI software that provides clinical decision support and influences clinical management decisions may require FDA clearance or approval. The determination turns on whether the software is intended to replace clinical judgment or merely provide information to a clinician who retains final authority. This distinction has significant product and governance implications. Features that are presented as clinical decision tools rather than informational aids may trigger a regulatory pathway requiring pre-market review.

Financial services AI operates under a layered structure of federal and state oversight. The Office of the Comptroller of the Currency, the Federal Reserve, and the FDIC have each issued guidance on model risk management, most notably in the Federal Reserve's SR 11-7, requiring banks to validate models, document assumptions, back-test predictions, and conduct independent model reviews. Critically, SR 11-7 applies to all models regardless of whether they use machine learning. What changes with AI is the complexity of the validation challenge: classical statistical models are interpretable; deep learning models are not, and regulators have signaled that black-box credit scoring raises fair lending concerns regardless of aggregate accuracy.

This matrix illustrates the complex landscape of federal oversight, mapping key regulatory bodies to specific industry verticals for compliance monitoring.

Employment AI tools that screen resumes, score candidates, or rank employees for promotion are among the most actively regulated AI domains in the United States. New York City's Local Law 144, effective in 2023, requires employers and employment agencies using automated employment decision tools to conduct annual bias audits by an independent auditor and to notify candidates that such tools are in use. Illinois, Maryland, and California have enacted or proposed similar requirements. The EEOC's AI and Algorithmic Fairness Initiative has made explicit that disparate impact analysis applies to AI-driven employment decisions. An employer cannot escape liability by attributing a discriminatory hiring pattern to an opaque algorithm; courts and regulators have consistently held that intent is not required to establish a disparate impact claim.

Government AI presents its own distinct regulatory structure. Federal agencies deploying AI in adjudicatory processes, determining benefits eligibility, enforcement priorities, and sentencing recommendations, face constitutional due process requirements in addition to the statutory frameworks that govern other sectors. The Administrative Procedure Act imposes requirements of reasoned explanation that intersect with AI explainability demands. If a federal agency relies on an AI model to deny a benefit or take an enforcement action, the agency's decision must be explainable in terms that can withstand judicial review. Several federal courts have already ruled that administrative determinations that rely on unexplained algorithmic scores without adequate opportunity for affected parties to challenge those scores fail to satisfy due process requirements.

4.6 Emerging AI Statutes

The European Union's AI Act represents the most comprehensive attempt to date to regulate AI as a category rather than through sector-specific lenses. Its risk-based framework classifies AI systems into four tiers: unacceptable-risk systems that are prohibited (e.g., real-time remote biometric surveillance in public spaces, social scoring), high-risk systems subject to extensive compliance obligations, limited-risk systems requiring transparency disclosures, and minimal-risk systems with no mandatory requirements. High-risk categories include AI used in critical infrastructure, educational assessment, employment decisions, access to essential services, law enforcement, migration, and the administration of justice. Organizations deploying high-risk systems must register them, conduct conformity assessments, maintain technical documentation, implement human oversight mechanisms, and ensure the systems are designed with appropriate accuracy, robustness, and cybersecurity safeguards.

Diagram 4.4: EU AI Act Risk Tier Pyramid

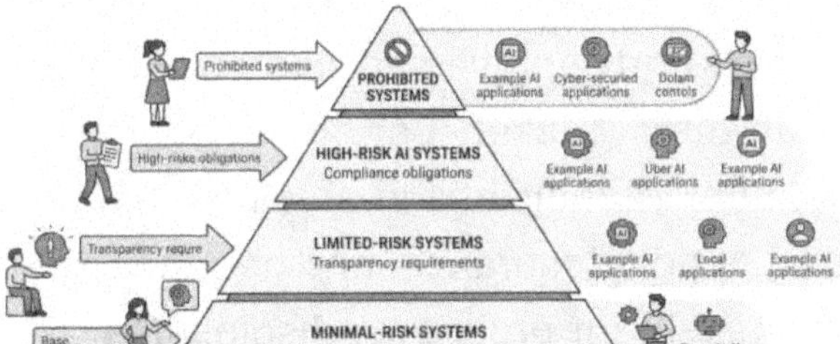

The EU AI Act classifies AI systems into four risk levels based on potential harm, defining prohibited uses and establishing compliance obligations for higher-risk applications.

In the United States, federal AI legislation has advanced incrementally. The AI Act of 2023 introduced requirements for federal agencies using AI in consequential decisions. Various National Defense

Authorization Acts have included AI governance provisions for defense and intelligence applications. The executive orders issued by successive administrations have directed federal agencies to adopt AI risk management frameworks aligned with NIST standards and to evaluate the civil rights implications of AI use in federal benefits and enforcement decisions. While federal comprehensive AI legislation had not yet passed as of this writing, the direction of regulatory travel is clear: disclosure, documentation, human oversight, and bias testing will be required. Organizations that build governance programs to meet those standards proactively will be better positioned than those that retrofit compliance after mandates are in place.

State-level AI legislation has proliferated since 2022. Colorado's AI Act, effective in 2026, applies to developers and deployers of high-risk AI systems defined as AI that makes or substantially influences consequential decisions in areas such as employment, housing, credit, insurance, and healthcare, and requires risk management programs, disclosures to consumers, and annual impact assessments. Similar frameworks are advancing in at least fifteen other states. The compliance challenge for organizations operating nationally is harmonization: when each state framework has slightly different definitions, thresholds, and documentation requirements, the governance program must be flexible enough to satisfy the most demanding applicable standard without becoming so burdensome that it impedes legitimate AI use. Maintaining a regulatory tracking process, monitoring pending legislation, and translating new requirements into program adjustments are operational responsibilities that must have a named owner.

4.7 Documentation Requirements

Across all legal frameworks that apply to AI, documentation is the linchpin of compliance. Regulators and courts assess compliance based on evidence of what the organization knew, when it knew it, what it decided, and what it did. An organization that conducted rigorous bias testing but documented none of it is legally indistinguishable from one that never tested at all. Documentation is not bureaucratic overhead; it is the instrument through which governance becomes auditable and defensible. Managers must treat documentation creation as a core deployment requirement, not as a retrospective record-keeping task.

The minimum documentation set for a legally defensible AI deployment includes: a system description sufficient for a non-technical regulator to understand what the system does and what decisions it influences; a data provenance record identifying the training datasets, their sources, and any known limitations; a bias and fairness testing log recording which tests were performed, what metrics were used, what thresholds were applied, and what the results were; a human oversight record showing how and when human reviewers interact with system outputs; an incident log recording any adverse events, system anomalies, or unexpected outputs; and a model version history tracking changes to the model over time. Each of these documents has a natural owner: data engineering, model development, operations, compliance, and each must be maintained as a living record, not a static artifact produced at deployment time.

Diagram 4.5: AI Documentation Portfolio

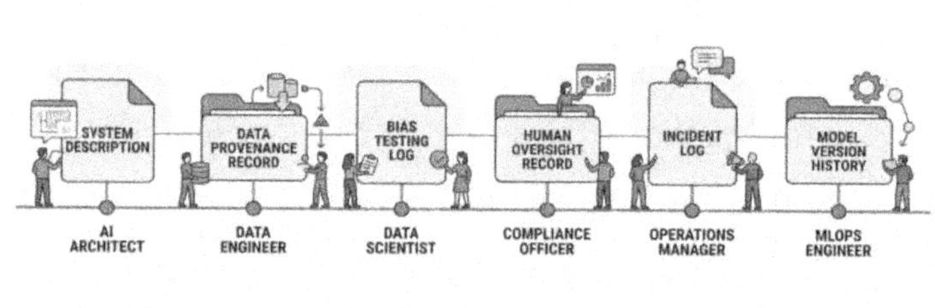

The AI Documentation Portfolio facilitates transparency, risk management, and regulatory compliance throughout the AI system lifecycle.

Retention periods for AI documentation depend on applicable law and the nature of the decisions being automated. EEOC regulations require that employment records relevant to a personnel action be retained for at least one year from the date of the action; some state laws require longer retention for automated employment decisions. FCRA-governed credit decisions require adverse action notice records to be kept for 25 months. HIPAA requires records of business associate agreements and privacy practices to be retained for six years. The practical approach is to define retention schedules by document type and applicable legal regime, review them annually as the legal landscape evolves, and implement the schedule in a records management system rather than leaving retention to individual file owners.

Model cards have emerged as a widely adopted documentation standard that addresses several legal documentation needs simultaneously. Originally developed at Google as a standardized model reporting format, a model card captures the model's intended use cases, performance metrics across demographic subgroups, evaluation data characteristics, ethical considerations, and known limitations. Many regulators and auditors have come to expect model cards as the baseline documentation artifact for AI systems. An organization that maintains current model cards for every production AI system has satisfied the descriptive documentation requirement common to virtually every regulatory framework and has positioned itself to respond quickly to any inquiry about what a system does and how it performs.

4.8 When to Engage Legal Counsel

Legal counsel is most valuable when engaged early in the AI deployment lifecycle, not when enforcement is imminent. The most common and most expensive mistake is treating legal review as a final checkpoint before deployment. By the time a system is fully built and trained, architectural decisions that would have been cheap to change at the design stage are expensive to reverse. Early legal engagement focuses on three questions: Which legal frameworks apply? What documentation must be produced before deployment? Are there design constraints, such as explainability requirements, human review mandates, or prohibited data uses, that must be built into the system architecture?

Specific triggers for legal counsel engagement include: any AI system that makes or substantially contributes to decisions about individuals in regulated domains (credit, employment, healthcare, housing, education, government benefits); any system that processes special categories of data (health information, biometric data, racial or ethnic data, financial data); any system intended for deployment in multiple jurisdictions with different legal requirements; any significant change to an existing AI system's scope, data inputs, or decision outputs; any regulatory inquiry, audit request, or litigation involving an AI system; and any vendor contract involving AI services where the organization is the deployer responsible for compliance.

Diagram 4.6: Legal Engagement Decision Tree

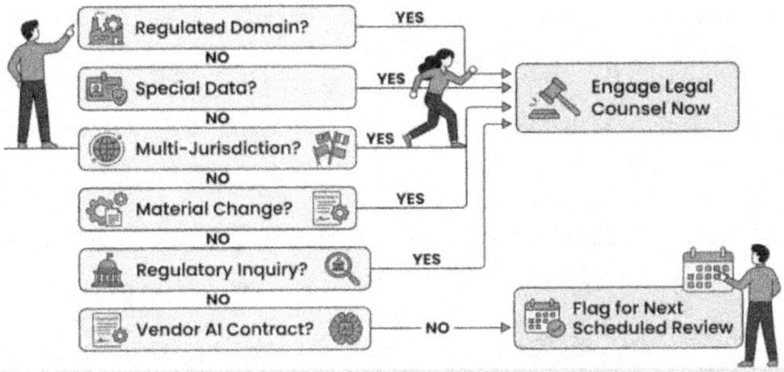

Systematic screening process to determine optimal legal counsel involvement.

Preparing effectively for legal counsel engagement means arriving with artifacts, not just questions. A manager who can hand legal counsel a system description document, a list of datasets used in training, a summary of what decisions the system influences, and a list of jurisdictions where it will operate enables a far more efficient and complete legal review than one who asks general questions about AI law. The manager's role is to translate technical and operational reality into terms that legal counsel can work with. The skill of being a good client of legal counsel is worth developing deliberately, as it reduces the cost of legal engagement and improves the quality of advice received.

Budget planning should include a standing allocation for legal counsel engagement on AI matters. In mature governance programs, legal review is not an emergency expense, but a planned cost embedded in the AI project lifecycle. The legal review for a new AI deployment in a regulated domain should be scoped and budgeted in the project plan, alongside model development, infrastructure, and testing. The legal review for existing systems should be included in the annual governance calendar alongside risk assessments and training cycles. Treating legal counsel as a planned resource rather than an emergency-response provider changes the economics of compliance from reactive crisis management to a predictable operating cost.

4.9 Preparing for Regulatory Inquiries

Regulatory inquiries about AI systems take several forms: formal investigative demands, audit requests from regulators with examination authority over the organization, and informal inquiries or questionnaires from agencies gathering information about industry practices. Each form requires a different response posture, but all share a common prerequisite: the ability to produce accurate, complete documentation on a defined timeline. Organizations that maintain contemporaneous documentation respond to inquiries in days or weeks. Organizations that must reconstruct documentation respond in months, at high legal cost and with the adverse inference that poor record-keeping implies poor practices.

Preparing for regulatory inquiries is a governance exercise, not a crisis response. It means defining in advance which systems are subject to regulatory oversight, who is responsible for maintaining documentation for each system, what the standard documentation set contains, and how quickly that documentation can be assembled and reviewed. It means conducting periodic internal tabletop exercises simulating an inquiry, pulling the documentation package, and identifying gaps before a regulator does. It means ensuring that key personnel, the model owner, the data engineer, the compliance lead, and the business line manager can accurately describe the system's operation and its governance controls without coaching.

Diagram 4.7: Regulatory Inquiry Response Workflow

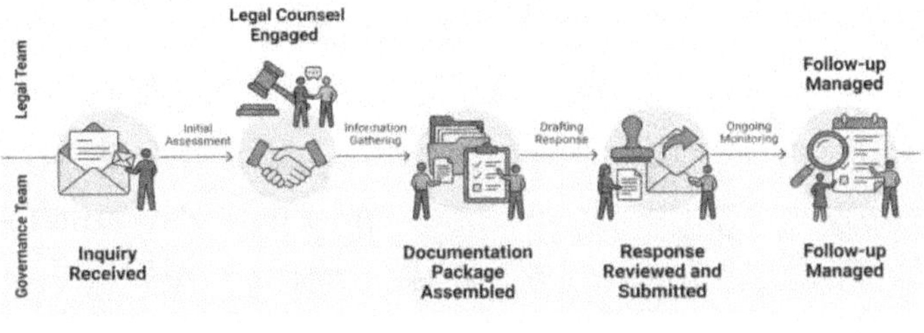

Caption: This regulatory inquiry responses wings to rerepulitor development the governance toaniment system.

When a regulatory inquiry does arrive, the response should be managed through legal counsel. The manager's role is to support counsel by providing accurate technical and operational information, not to respond directly to regulators without legal coordination. The information provided should be accurate, complete, and limited to what has been requested. Speculation, estimation, and volunteering information beyond the scope of the inquiry are common mistakes in regulatory responses that create additional exposure. The most defensible response is grounded in contemporaneous documentation maintained as a matter of routine governance practice, not assembled retroactively to meet an inquiry deadline.

Post-inquiry analysis is a governance input that is frequently overlooked. After every regulatory inquiry, the governance team should conduct a structured debrief: what documentation was requested, what gaps were identified when assembling the documentation package, how long it took to produce a complete response, and what control improvements would close the gaps for the next inquiry? This analysis should be documented and incorporated into the governance improvement backlog. Organizations that systematically learn from regulatory inquiries improve their governance posture over time; organizations that treat each inquiry as a discrete crisis learn nothing and incur the same costs repeatedly.

4.10 Manager's Checklist — Legal Framework

Before deploying any AI system that makes or influences decisions about individuals, complete the following steps. Map the system to applicable legal frameworks by industry vertical, data type, and jurisdiction. Identify which categories of personal data the system processes and confirm the lawful basis for each processing activity. Determine whether the system falls into any high-risk category under the EU AI Act or applicable state AI laws and document that determination. Confirm that automated decision-making outputs in regulated domains route through a human review step and that the review is documented. Engage legal counsel at the design stage if the system operates in the areas of credit, employment, healthcare, housing, government benefits, or education.

Build the minimum documentation set system description, data provenance record, bias testing log, human oversight record, incident log, model version history before go-live, not after. Define retention schedules for all AI documentation in accordance with applicable legal requirements and record them in the records management system. Identify which regulatory bodies have oversight authority over the organization's AI systems and review their most recent guidance on AI. Conduct an annual tabletop exercise simulating a regulatory inquiry for at least one production AI system and document the gaps identified. Establish a named point of contact in legal counsel for AI matters so that escalation paths are clear before an inquiry arrives. Designate a named owner for regulatory tracking — monitoring pending AI legislation and translating new requirements into governance adjustments — and confirm that this responsibility appears in their role definition and performance plan.

4.11   Key Takeaways

Legal compliance for AI is a management responsibility that cannot be delegated solely to lawyers or technologists. The legal landscape is layered with federal sector laws, state privacy statutes, emerging AI-specific legislation, and international frameworks that apply simultaneously, and the most restrictive applicable requirement sets the compliance floor. Documentation is the operational instrument of legal compliance: it is what transforms governance intentions into auditable evidence. Early legal engagement is cheaper and more effective than emergency response. The EU AI Act's risk-tier framework and the proliferating state-level AI statutes are converging toward common requirements: transparency, human oversight, fairness testing, and documented risk management. Managers who build governance programs oriented to those requirements will find that compliance with specific statutes, when they arrive, requires refinement rather than reconstruction. Post-inquiry analysis is a discipline that converts regulatory encounters from pure cost into governance improvement. The legal framework does not reward good intentions — it rewards demonstrable practices.

5 Implementing Governance and Risk Management

5.1 Opening Scenario

A federal agency's technology division had produced a 47-page, well-organized AI governance policy document, approved by the agency's IT governance board. Twelve months after approval, a compliance review found that fewer than a third of production AI systems had been registered in the inventory required by the policy. None of the program's risk assessments had been completed on schedule. Three of the four governance roles defined in the policy had no named incumbents. The training program described in the policy had never been designed. The policy had been written as an end product rather than a starting point. No one had answered the harder questions: Who will do each task? By when? With what resources? The failure was not one of policy quality; the document was technically sound. It was a failure of implementation discipline. Governance that exists only on paper manages no risk.

Organizations that have navigated this failure mode once tend to make the same correction: they shift from treating governance as a documentation project to treating it as an operational program. Documentation is necessary — policies, procedures, charters, and plans are real governance artifacts. But they are necessary conditions for governance, not sufficient ones. Sufficient governance requires that someone is executing those documents in daily operations: conducting risk assessments on schedule, reviewing monitoring alerts, completing training, and processing changes through the right gates. The shift from a documentation project to an operational program is the central challenge in implementing AI governance.

5.2 Why Implementation Matters

AI governance programs fail most often not because their policies are wrong but because their implementation plans are missing. A governance framework is a theory for managing risk. An implementation plan is the proof that the theory will be enacted. Without a concrete, sequenced, resourced plan, every governance initiative encounters the same fate: initial enthusiasm, diffuse accountability, and eventual stagnation. The risk does not go away when governance stagnates; it accumulates.

The challenge of implementation is that governance work competes with operational work for the same limited pool of time, budget, and attention. In that competition, operational urgencies almost always win unless governance activities are embedded into routine operational processes rather than maintained as parallel programs. The manager who treats governance as a separate track that runs alongside real work will lose. The manager who integrates governance into how work gets done, into project gates, into change control, into vendor onboarding, and into incident response will sustain it.

This chapter provides a step-by-step implementation roadmap grounded in the realities of enterprise IT organizations. It does not assume unlimited resources or a blank-slate environment. It assumes that the organization has existing IT processes, existing compliance functions, existing staff with day jobs, and competing priorities. The approach is incremental: establish a baseline, identify the highest-priority risks, implement the most valuable controls first, and build program maturity over successive cycles. The manager who finishes this chapter should be able to open a calendar and schedule the first governance milestone, not because the problems are simple, but because the first steps are clear.

5.3 The Implementation Roadmap

A governance implementation roadmap serves two functions simultaneously: it sequences work logically so that foundational capabilities are built before advanced ones, and it creates visible progress milestones that sustain organizational commitment over the multi-year lifecycle of program buildout. A roadmap without milestones is a timeline. A roadmap without sequencing is a backlog. Both are insufficient. The roadmap described here organizes implementation across three phases: foundation, operationalization, and continuous improvement.

The foundation phase, typically spanning the first 90 days, focuses on three deliverables: an AI system inventory, an initial risk assessment, and a governance structure with named roles. The AI system inventory is the critical first artifact; you cannot manage what you have not cataloged. The inventory should record each AI system in production or active development, the business function it supports, the decisions it influences, the data it processes, and the team responsible for it. The risk assessment scores each inventoried system against a consistent set of risk factors: consequence severity, decision autonomy, data sensitivity, regulatory exposure, and population scale. The governance structure assigns the roles defined in the organization's governance charter to actual named individuals and communicates those assignments formally.

Diagram 5.1: Three-Phase Implementation Roadmap

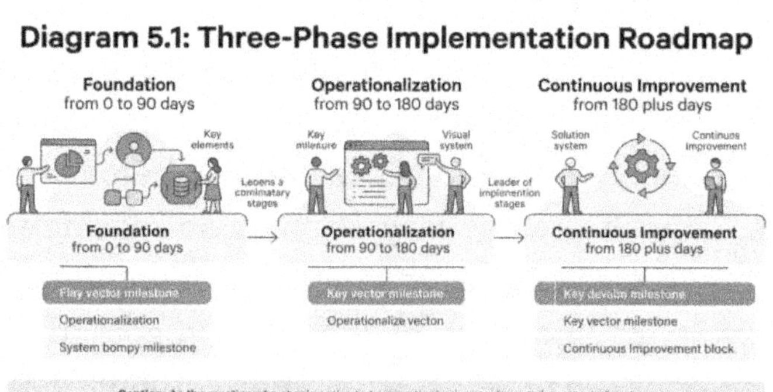

The operationalization phase, spanning roughly months three through six, translates the risk assessment into active controls. Each high-risk system identified in the foundation phase receives a control plan: a specific set of technical and procedural controls mapped to the risk factors that drove its high-risk rating. Policies are drafted, reviewed, approved, and published. The training program is designed and delivered to the first cohort of personnel. Monitoring mechanisms, dashboards, alerts, and review cadences are configured. The operationalization phase is when governance becomes tangible to the people doing the work rather than only visible to those managing the program.

The continuous improvement phase is indefinite. It is the operational steady state of a mature governance program: regular risk reassessment as systems evolve, systematic review of control effectiveness, structured incorporation of lessons from incidents and near-misses, and deliberate advancement of maturity on identified capability gaps. The continuous improvement phase is distinguished from the foundation and operationalization phases not by its activities but by its rhythm, which operates on predictable cadences rather than project timelines. Governance maturity under continuous improvement is measured in years: the program that completes five annual assessment cycles has institutional memory, calibrated controls, and practitioner expertise that a program in its first year does not.

5.4 Risk Identification

Risk identification for AI systems requires a structured approach that is both comprehensive, covering all meaningful risk categories, and repeatable, producing consistent results across different assessors and different systems. The NIST AI Risk Management Framework's four primary risk categories provide a useful organizing structure: trustworthiness risks (bias, explainability, reliability), operational risks (system failures, integration failures, dependency risks), security risks (adversarial attacks, data poisoning, model theft), and legal and compliance risks (regulatory violations, contractual obligations, litigation exposure). Each category requires a different assessment technique and involves different subject matter experts.

Risk identification workshops are the most effective way to surface risks that lie in the gap between technical teams and business functions. The workshop brings together the model developer, the data engineer, the business process owner, the compliance lead, and the operational staff who interact with system outputs. Each group sees risks the others do not: the business process owner understands what happens when the system produces an incorrect output in a high-stakes situation; the data engineer understands the quality limitations of the training data; the compliance lead understands the regulatory exposure; the operational staff understands the practical workarounds that have evolved around system limitations. The facilitator's job is to surface these perspectives systematically and document them.

Diagram 5.2: AI Risk Identification Workshop Structure

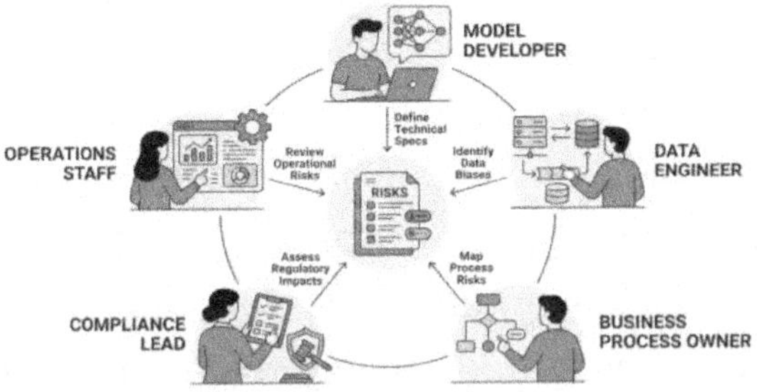

Stakeholders collaborate to identify, categorize, and document potential AI risks within the workshop framework.

The output of risk identification is a populated risk register. Each entry in the register should record: a concise risk statement describing what could go wrong and what the consequence would be; the system or systems affected; the risk category; an inherent risk rating using a consistent severity and likelihood scale; the controls currently in place; the residual risk rating; the risk owner; and the next review date. The risk register is not a static document; it should be updated whenever a system changes materially, whenever a new risk is identified, and on a scheduled review cycle aligned with the governance program's cadence.

Scoring consistency across risk assessments is a practical challenge that deserves explicit attention. When the same system is assessed by different people six months apart and produces different scores, the risk register loses its value as a comparison tool. Scoring rubrics that define each point on the severity and likelihood scales, with example situations, are the primary mechanism for ensuring consistency. Calibration sessions, in which two or more assessors independently score the same system and then compare and reconcile their scores, are valuable when a new governance program is establishing its baseline. The goal is not to eliminate judgment but to ensure that different assessors applying the same rubric reach similar conclusions for similar risk profiles.

5.5 Control Selection

Control selection is the process of choosing which risk mitigation mechanisms to implement for each identified risk. Not every risk needs technical control. Not every control needs to be custom-built. The manager's task is to match each material risk to an appropriate control from a range of options: technical controls built into the system architecture, procedural controls embedded in operational workflows, contractual controls governing vendor obligations, and detective controls that surface problems after they occur. Two questions should govern the selection: Does this control reduce the risk it addresses? Is the cost of implementing and maintaining this control proportionate to the risk it addresses?

The NIST SP 800-53 control catalog provides a comprehensive baseline applicable to federal systems and widely adopted in commercial contexts. AI-specific control guidance from the NIST AI RMF Playbook maps governance practices to each of the framework's four core functions: Govern, Map, Measure, and Manage. ISACA's AI Audit Framework provides additional control taxonomies oriented toward assurance needs. The practical approach is to start with the organization's existing IT control framework, whichever baseline is already in use, and identify the incremental controls needed to address AI-specific risks not covered by standard IT controls.

Diagram 5.3: Control Selection Decision Matrix

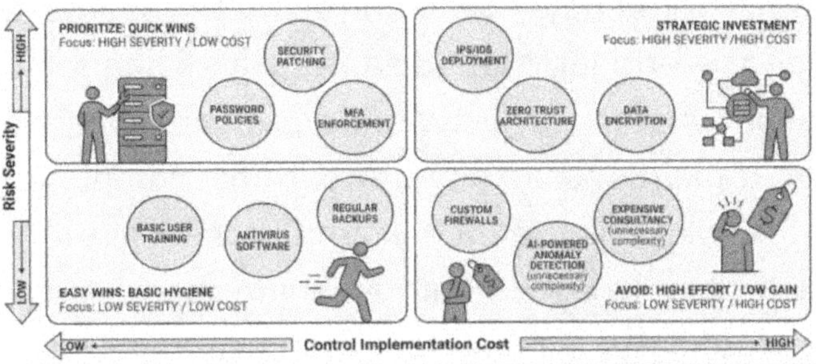

The Control Selection Decision Matrix provides an objective framework for prioritizing risk mitigation strategies based on severity and cost.

AI-specific controls that frequently appear in well-designed governance programs include model performance monitoring with defined thresholds and automated alerts, human review gates for high-stakes automated decisions, bias testing on representative evaluation datasets before deployment and at defined post-deployment intervals, data quality scoring for model inputs with rejection or flagging of inputs that fall below thresholds, adversarial testing protocols that simulate attempts to manipulate model outputs, and model versioning and rollback capability that enables rapid recovery from degraded model performance. Each of these controls has design requirements that should be specified before development begins, not bolted on after deployment.

The control selection process should produce a control matrix: a document that maps each identified risk to one or more controls, records the control owner, specifies how the control is tested, and indicates the testing frequency. The control matrix is the primary evidence artifact that regulators and auditors use to assess whether a governance program addresses the risks it claims to manage. A current, complete control matrix linked to evidence of control testing is among the most valuable documents an organization can produce in response to a regulatory inquiry. Organizations that maintain it as a living document spend hours on regulatory inquiries; those that do not spend weeks.

5.6 Policy Drafting

Effective AI governance policies are operational documents, not position statements. They tell specific people what to do in specific situations, not general principles that everyone endorses and no one acts on. The test of a policy is operational: could a new employee, reading only this document, understand what they are required to do, when they are required to do it, and who they should contact if they have a question? If not, the policy requires revision before publication.

Policy drafting begins with scope definition. Who does this policy apply to? What systems or processes does it govern? What time periods or lifecycle stages does it cover? Scope ambiguity is the most common cause of policy non-compliance. When people are unsure whether a policy applies to their situation, they default to the interpretation that requires the least additional work. A well-scoped policy leaves no room for that ambiguity. Following the scope definition, the policy should specify requirements, the specific things that must be done, clearly distinguished from guidance, which describes how requirements might be met without mandating a specific approach. Requirements use language such as shall, must, and is required. Guidance uses language such as should, may, and is recommended.

Diagram 5.4: Policy Drafting Lifecycle

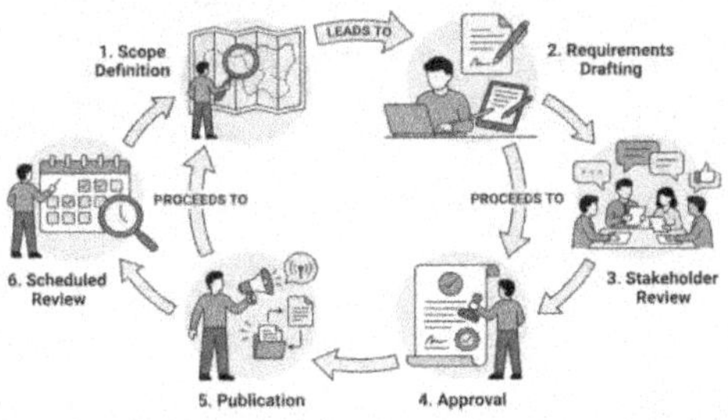

This iterative cycle ensures structured development, stakeholder alignment, and continuous refinement for effective organizational policies.

The approval and review cycle for AI governance policies should, at a minimum, involve the technical lead responsible for the systems covered, the business process owner whose operations are affected, the legal or compliance function with oversight responsibility, and the governance body with authority to approve. Policies should have a stated review date, typically annual, and a named policy owner who is responsible for triggering the review. Policies that are not reviewed become artifacts of a past governance moment rather than current operational requirements. The policy owner role is not ceremonial; it carries a specific accountability for currency and accuracy.

A minimum viable AI governance policy set for most enterprise organizations includes: an AI acceptable use policy establishing permitted and prohibited uses of AI tools; a model development and documentation policy specifying what artifacts must be produced for each AI system; a data governance policy for AI addressing training data sourcing, quality standards, and retention; a third-party AI policy governing vendor AI services; and an AI incident response policy defining how adverse events involving AI systems are reported, investigated, and remediated. These five policies address the majority of the risk surface for most AI deployments and represent the policy foundation that more specific controls and procedures should reference.

5.7 Training and Awareness

Governance controls and policies produce no risk reduction without the people who operate under them understanding what they require and why. Training is the mechanism through which written policy becomes operational behavior. The training program for AI governance must be role-differentiated: the AI developer needs to understand model documentation requirements and bias testing procedures; the business analyst needs to understand how to interpret model outputs and recognize signs of performance degradation; the senior manager needs to understand the governance framework, their decision rights, and their escalation obligations; the legal and compliance professional needs to understand AI-specific risk categories and documentation standards. One-size-fits-all training satisfies a compliance checkbox while achieving little behavioral change.

Training delivery should combine initial certification training required for any personnel taking on an AI governance role, with ongoing awareness activities that maintain engagement between formal training cycles. Awareness activities might include brief case studies of governance incidents shared at team meetings, periodic reviews of the program's risk register with the relevant operational team, tabletop exercises that walk through a simulated incident or regulatory inquiry, and updates on regulatory developments affecting the organization's governance obligations. The goal of awareness activities is to keep governance visible and relevant in the day-to-day work environment rather than treating it as an annual compliance obligation.

Training completion rates and assessment scores should be tracked and reported to the governance committee on a defined schedule. Departments or teams with consistently low completion rates may indicate understaffing, unclear role assignments, or insufficient management sponsorship for governance activities. These are management problems, not training problems, and they require managerial intervention rather than more training content. The governance program should not accept low compliance as a baseline; it should investigate the causes and address them. Completion rates below 80 percent in any team with governance responsibilities are a governance finding, not an administrative inconvenience.

5.8 Continuous Monitoring

AI systems degrade in ways that are not always visible from outside the model. Training data becomes stale as the world changes. Model performance drifts as the input distribution shifts away from the training distribution. Edge cases that were rare during testing become common as deployment scales. Biases that were acceptable at low decision volumes become material at high volumes. Continuous monitoring is the control mechanism that makes these degradation patterns visible before they produce significant harm or regulatory exposure.

A monitoring program for AI systems should cover at least four dimensions: performance metrics (accuracy, precision, recall, or domain-specific performance measures), fairness metrics (disparate impact, equalized odds, or other metrics appropriate to the decision context), operational metrics (system availability, response latency, error rates, input data quality), and business outcome metrics (decision rates, appeal rates, manual override rates, downstream outcome tracking). Each dimension requires defined thresholds that trigger review when crossed. The thresholds are not arbitrary; they should be set based on the severity of the consequences of the decisions the system influences and on the maximum acceptable degradation before re-evaluation is required.

Diagram 5.5: Continuous Monitoring Dashboard Layout

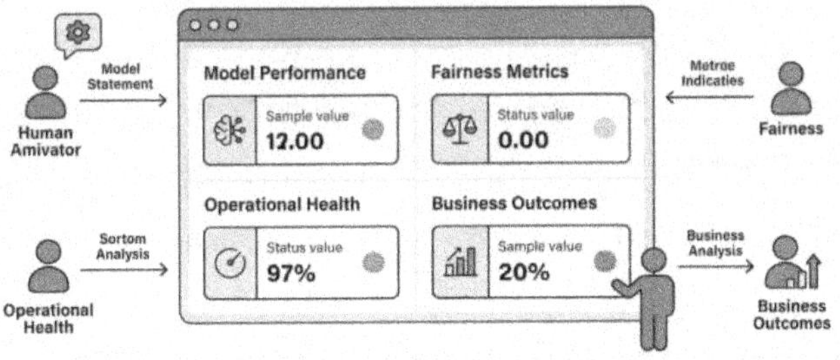

The executive Continuous monitoring dashboard layout: sr analysize bioine model performance a operational health, and hunkings way to sencise a models.

Monitoring results must connect to response procedures. A dashboard showing a model performance alert that no one responds to is a governance failure worse than no monitoring at all. It creates a record that the organization knew of a problem and did nothing. Response procedures for monitoring alerts should be documented, tested, and assigned to named roles. When a threshold is crossed, the response procedure should specify who is notified, what investigation is required, what immediate mitigations are available (including suspending the system if necessary), and what documentation must be produced. These procedures should be tested in regular drills, not only during real events.

The monitoring infrastructure for AI systems should be treated as a production system, not an afterthought. Monitoring data should be retained long enough to support regulatory inquiries and incident investigations, typically for at least two years, and longer in regulated decision-making contexts. The monitoring system should be designed to survive model updates without losing historical metric continuity. A monitoring architecture that starts fresh with each retraining event cannot support trend analysis or regulatory retrospectives. Incident response procedures should specify the maximum acceptable time between alert trigger and first-line response. That timeline should be enforced through escalation protocols if the initial response is not timely.

5.9 Overcoming Implementation Barriers

The barriers to implementing AI governance are predictable and recurring. Understanding them in advance allows managers to address them proactively rather than being surprised by them when they materialize. The most common barriers fall into four categories: resource constraints, organizational resistance, capability gaps, and scope disputes.

Resource constraints are the most frequently cited barrier and the most commonly used excuse for inaction. Governance does require resources — time, budget, personnel — but well-designed governance does not require large dedicated teams. In most organizations, the most resource-efficient approach is to integrate governance activities into existing roles and processes rather than creating a standalone governance function. The change control team can incorporate AI change review into existing change advisory board processes. The compliance team can incorporate AI risk into existing risk assessment cycles. The vendor management team can incorporate AI supplier assessments into existing vendor review processes. Integration reduces marginal resource requirements while increasing sustainability, because governance activities endure turnover and budget cycles better when embedded in operational functions than when they rely on a dedicated team.

Diagram 5.6: Implementation Barrier Response Map

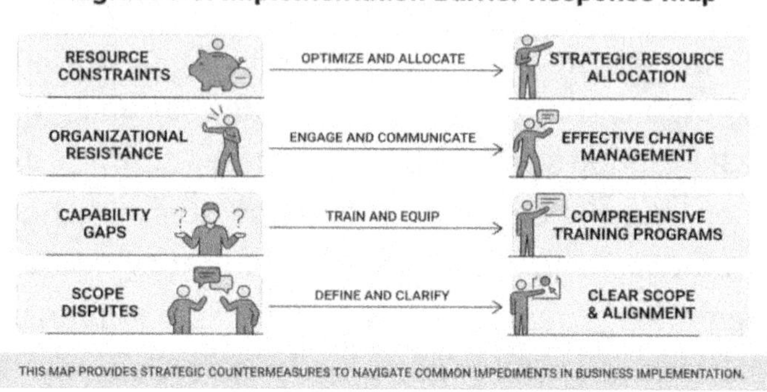

Organizational resistance typically emerges from AI development and business teams that perceive governance as a slowdown without a corresponding benefit. The most effective response to this resistance is not to argue for governance in the abstract but to demonstrate its practical value concretely: show how documentation requirements that feel burdensome during development prevent expensive remediation work later; show how risk assessments that take a few hours prevent incidents that take weeks to resolve; show how vendor oversight processes that add days to procurement prevent the integration of AI tools that create legal liability. Governance that is experienced as purely overhead will be resisted. Governance that visibly prevents real problems will be accepted and eventually championed.

Capability gaps, situations where the governance program requires skills or expertise that the organization does not currently have, are the barrier that takes the longest to close. Bias testing, model explainability assessment, and AI-specific risk analysis require technical expertise that many IT governance functions do not yet have in-house. The near-term response is to contract that capability from consulting partners or engage vendors who provide it as a service. The longer-term response is to build internal capability through training, hiring, and partnerships with the organization's data science and AI development teams, which hold the technical knowledge that governance functions need. Capability gaps should be tracked as part of the governance program's maturity assessment, with a named owner and a target closure date for each gap.

Scope disputes, disagreements about whether a particular system or activity falls within the governance program's scope, are best resolved through clear, documented scope definitions approved by the governance committee. When the governance program's scope is specified with enough precision that reasonable people applying it reach the same conclusion for a given system, disputes are rare. When the scope is defined in general terms that require interpretation, disputes are frequent and recurrent. The investment in precise scope definition pays dividends in reduced governance overhead over the program's lifecycle. Every scope dispute that is not resolved through clear documentation will recur.

5.10 Manager's Checklist — Implementation

To launch a governance implementation program, complete the following foundational steps in the first 90 days. Conduct an AI system inventory using a consistent template; assign each inventoried system to a named owner. Complete an initial risk assessment for each inventoried system using a consistent scoring rubric; document the methodology and results. Formally assign named individuals to each governance role defined in the governance charter; communicate the assignments and confirm acceptance. Identify the three to five highest-risk systems from the inventory and develop a control plan for each. Engage legal counsel to confirm the applicable legal frameworks for each high-risk system and identify any documentation gaps.

In the operationalization phase, draft and approve policies covering the highest-risk process areas first; publish with named policy owners and review dates. Design and deliver role-differentiated training to all personnel with governance responsibilities, track and report completion rates by role. Configure monitoring dashboards for all high-risk production systems with defined alert thresholds and documented response procedures; test the response procedures before relying on them. Establish a governance cadence with monthly operational reviews, quarterly risk register reviews, annual program assessments, and calendar the first occurrence of each. Conduct a tabletop exercise simulating either a regulatory inquiry or a model performance incident and document findings and remediation actions. Build the governance metrics dashboard inventory coverage, risk-assessment currency, training completion, alert response times, and control testing coverage, and begin reporting to the governance committee monthly.

5.11   Measuring Program Effectiveness

A governance program that does not measure its own effectiveness cannot demonstrate value, identify underperformance, or make evidence-based decisions about where to invest improvement effort. Measurement is not a reporting exercise; it is the feedback mechanism through which a governance program learns whether its controls are working. Managers who understand what to measure and how to interpret results will build programs that improve continuously. Managers who treat measurement as a reporting obligation will produce metrics that satisfy audit requirements but do not inform decisions.

The most useful governance metrics measure outcomes, not activities. Activity metrics, such as the number of risk assessments completed, the number of training hours delivered, and the number of policies published, confirm that governance activities were executed. They do not confirm that those activities reduced risk. Outcome metrics address risk reduction directly: the number of model performance incidents in a reporting period, the mean time to detect and respond to monitoring alerts, the number of regulatory findings in the most recent examination, and the rate of adverse events attributed to AI system failures. A program that shows declining incident rates and faster detection times across successive reporting periods provides evidence that its controls are working. A program that only shows activity counts provides evidence that its staff is busy.

Diagram 5.7: Governance Program Effectiveness Metrics

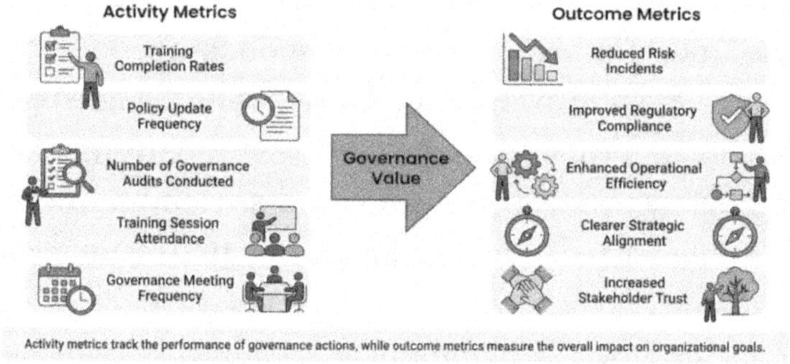

Activity metrics track the performance of governance actions, while outcome metrics measure the overall impact on organizational goals.

Reporting governance metrics to the governance committee and to senior management should be structured as a narrative, not a table dump. A metrics report that presents 20 numbers without context or analysis provides no more decision support than no metrics at all. The governance report should answer three questions for its audience: Is the program executing as designed? Are the controls working? What is changing, and what does it mean? The answer to each question should be supported by data, accompanied by the governance team's interpretation, and, where action is needed, a specific recommendation. Governance reports that ask leadership to draw their own conclusions from raw data are not governance reports; they are data exports. The governance function's job is to analyze, interpret, and recommend, not merely to tabulate.

Program effectiveness measurement also informs resource allocation decisions. When the governance metrics show that a particular control domain, say, vendor oversight, has high findings rates and slow remediation times relative to other domains, that is evidence that the domain is under-resourced or under-designed. It is the basis for a budget conversation grounded in evidence of risk rather than managerial intuition. Conversely, when the control domain consistently produces clean results across multiple cycles, that may be evidence that resources can be redistributed to higher-risk areas without increasing overall risk exposure. A governance program that tracks effectiveness metrics over time acquires the data needed to make these decisions rationally. One that does not will make them do so by intuition, sometimes correctly, but without the institutional knowledge to explain why or to repeat the success.

5.12   Key Takeaways

Governance programs succeed or fail at the implementation stage, not the design stage. A roadmap that sequences foundation work before operationalization and operationalization before continuous improvement gives programs the best chance of building durable capability rather than producing documents that no one acts on. Risk identification, control selection, policy drafting, training, and monitoring are distinct management tasks with defined outputs, and all five must be completed before a governance program can be considered operational. The barriers to implementation are predictable: resource constraints, organizational resistance, capability gaps, and scope disputes. Each barrier has a known countermeasure. Managers who anticipate those barriers and plan for them will sustain governance programs that continue to function when they are needed most. Governance integrated into existing operational processes is more sustainable than governance that operates as a parallel program. The continuous improvement phase, indefinite, cadence-driven, metrics-tracked, is what separates programs that remain current from programs that calcify. Measuring outcomes rather than activities is the discipline that transforms reporting into learning.

6 Improving IT Governance

6.1 Opening Scenario

A logistics company's IT organization ran a competent change management process for its legacy systems, with seven-day review cycles, documented rollback plans, and post-implementation reviews. When the team deployed its first machine-learning-powered route-optimization system, the change advisory board processed the deployment using the same form it used for software patches. No one asked how the model had been validated. No one documented what the training data represented. When the model was retrained six weeks after deployment to incorporate new traffic data, that retraining was not submitted to the change board at all; the data science team considered it a model update rather than a system change. Three months later, the model began producing suboptimal routes during peak periods. The root cause was a data quality issue introduced in the retraining dataset. The change board had no record of the retraining and therefore no audit trail to support the investigation. The incident took eleven days to diagnose because no one could trace the sequence of changes. The company's legacy change process had not been wrong; it had never been extended to cover the specific characteristics of AI system changes.

The lesson from that incident extended beyond the immediate fix. After the investigation, the logistics company's IT governance lead conducted a broader review and found five other active AI systems that were similarly invisible to existing governance processes: not maliciously hidden, but simply not within the scope of processes designed before AI systems existed. Each of those systems was making consequential operational decisions without change-control coverage, a configuration record, or monitoring thresholds. The company's governance infrastructure was perfectly adequate for the systems it had been designed to govern. It had not kept pace with the technology it was now responsible for.

6.2 Why IT Governance Improvement Matters

AI governance does not exist in a vacuum. Every AI system runs on an IT infrastructure governed by existing IT processes, change management, configuration management, vendor oversight, and incident response. When those processes are not extended to account for AI-specific characteristics, governance gaps appear at the seams: places where AI systems fall through the cracks of existing controls because they were designed for a different technology paradigm. Improving IT governance for AI means systematically identifying gaps and closing them, not building a parallel governance apparatus that duplicates the existing one.

The economic case for extending existing IT governance processes rather than creating new ones is compelling. Every standalone governance program competes with other priorities for budget and attention. A governance requirement embedded in an existing process that already has organizational legitimacy, staffing, and tools costs far less to sustain than an equivalent standalone requirement. The change advisory board, which already reviews system changes, is the right venue for reviewing AI model changes; it needs to be extended, not replaced. The configuration management database that already tracks system components is the right system of record for AI model versions; it needs additional data fields, not a separate system.

Frameworks such as COBIT and ITIL provide structured vocabularies and process models that most enterprise IT organizations already use in some form. Aligning AI governance with these frameworks accelerates adoption by connecting new requirements to familiar concepts, leveraging existing tooling and reporting infrastructure, and making AI governance comprehensible to auditors and regulators accustomed to assessing IT governance using these frameworks. The manager who can describe AI governance in COBIT or ITIL terms is a far more effective communicator with audit and compliance functions than one who presents AI governance as an entirely novel program requiring entirely novel frameworks.

6.3 IT Governance Foundations

IT governance is the set of processes, structures, and accountability mechanisms through which an organization ensures that its IT investments deliver value and that IT-related risks are managed to acceptable levels. The foundational elements of IT governance strategy alignment, value delivery, risk management, resource management, and performance measurement apply to AI systems in the same way they apply to enterprise applications, infrastructure, and data platforms. An AI system is an IT asset. It consumes resources. It delivers business outcomes. It creates risks. It needs to be governed with the same discipline applied to other IT assets, adapted for the specific characteristics that distinguish AI from conventional software.

The most important characteristic that distinguishes AI systems from other IT assets for governance purposes is behavioral non-determinism combined with continuous change. A traditional application executes instructions written by humans; its behavior changes only when its code is changed. An AI model generates outputs from learned patterns; its effective behavior can change without any change to its code, through drift in the input data distribution, changes in the model's operating environment, or the accumulation of edge cases that the training data did not represent. This behavioral non-determinism means that AI systems require a continuous operational oversight mechanism that traditional IT governance processes often lack.

Diagram 6.1: Traditional IT Governance vs. AI Governance Overlap

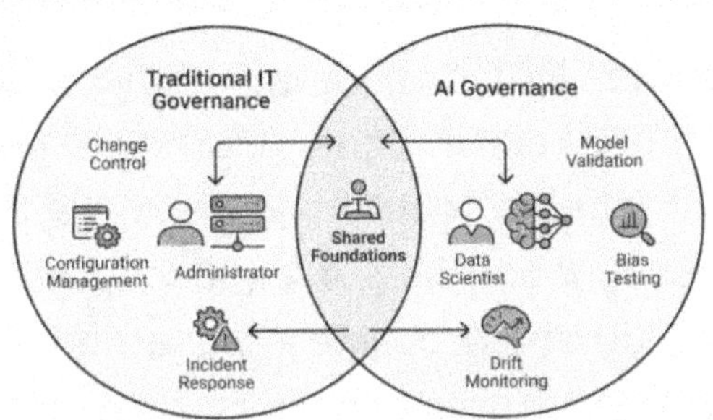

Integrating specialized AI oversight with established IT governance ensures responsible, efficient technology management.

The practical implication is that extending IT governance to AI requires adding oversight mechanisms designed for continuous behavioral change, not just for discrete system changes. A model retrained on new data is not the same system it was before, even if its code is identical. A model operating on inputs that have a shifted distribution is not performing the same function it was validated for, even if no one has touched the system. IT governance processes must be updated to recognize these realities and to require the operational reviews that detect and respond to them. The governance function that does not update its understanding of what constitutes a system change will inevitably miss material changes to AI systems.

Strategy alignment, ensuring that AI investments are aligned with the organizational mission and evaluated against strategic objectives, is an IT governance function that becomes more critical as AI adoption scales. Organizations that deploy AI systems without asking whether the deployment advances a defined strategic objective often find, on review, that they are maintaining a portfolio of AI experiments that consume operational resources without delivering consistent value. The governance function should require that every AI system have a documented business case tied to a strategic objective, a defined set of success metrics, and periodic reviews to assess whether those metrics are met. Systems that do not meet their defined objectives should be candidates for retirement rather than indefinite maintenance.

6.4 COBIT and ITIL Alignment

COBIT 2019 organizes IT governance and management objectives into six domains: Evaluate, Direct, and Monitor at the governance level; Align, Plan, and Organize; Build, Acquire, and Implement; Deliver, Service, and Support; and Monitor, Evaluate, and Assess at the management level. Each domain contains specific practice objectives with defined maturity indicators. AI governance activities map naturally to COBIT's structure: AI strategy and portfolio decisions belong in the Align, Plan, and Organize domain; AI system development and procurement in Build, Acquire, and Implement; AI operational management in Deliver, Service, and Support; and AI audit and assurance in

Monitor, Evaluate, and Assess. Mapping AI governance requirements to the COBIT structure enables organizations to assess their AI governance maturity using COBIT's capability levels and to track improvement against a recognized standard.

ITIL 4 provides a service management perspective organized around the Service Value System, a model showing how all components of an IT organization work together to create value for stakeholders. ITIL's service management practices most relevant to AI governance include change enablement, problem management, service configuration management, and information security management. Each of these practices needs to be updated to account for AI-specific characteristics: change enablement must address model retraining events, problem management must address model drift as a root cause category, configuration management must include model artifacts as configuration items, and information security must address AI-specific attack vectors such as adversarial examples and data poisoning.

Diagram 6.2: COBIT Domain Mapping for AI Governance

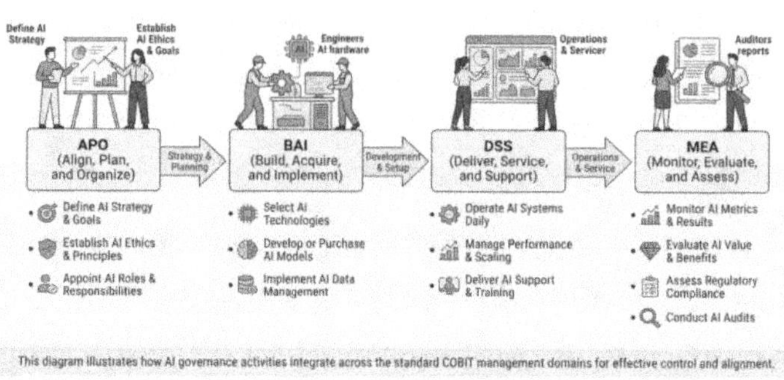

Organizations that have achieved COBIT or ITIL maturity in their general IT governance will find that extending these frameworks to AI is more an exercise in scope expansion and practice adaptation than one of building from scratch. The governance vocabulary is familiar, the committee structures are in place, and the reporting mechanisms are in place. What is needed is a deliberate assessment of which existing practices apply directly to AI, which need adaptation, and which AI-specific requirements have no analog in the existing framework and need to be added as new practices. That gap analysis, conducted systematically, produces a practical AI governance improvement roadmap grounded in institutional capability.

Maturity assessment against COBIT's capability levels provides a structured language for communicating the status of governance programs to executive leadership and audit functions. A program at

Capability Level 1 — Performed — executes governance activities but without consistent standards or documented outcomes.

Capability Level 2 — Managed — executes activities with defined plans, resources, and quality standards.

Capability Level 3 — Established — executes activities using defined organizational processes.

Most enterprise IT governance programs for AI should target Capability Level 2 as an initial goal and Capability Level 3 as the mature-state target. Targeting Capability Level 5 in the first year is not a governance ambition — it is a distraction from building the foundational practices that every higher capability level depends on.

6.5 Change Control for AI

Change control for AI systems must address a broader range of change types than traditional IT change control. In traditional change management, the primary types of change are code releases, infrastructure changes, and configuration changes, all of which involve deliberate human decisions to modify a system. AI systems introduce two additional change types that have no traditional analog: model retraining events, in which the model's learned parameters are updated without changing its code; and data distribution shifts, in which the statistical properties of the inputs the model receives change over time without any deliberate action by the organization.

Model retraining events should be subject to formal change control regardless of whether the model code changes. A retraining event changes the model's behavior by updating its parameters. That behavioral change carries the same risk profile as a code change: it may introduce new biases, reduce performance on previously well-handled subpopulations, or change decision outputs in ways that have regulatory or operational significance. The change review for a retraining event should include: confirmation that the retraining dataset has been quality-assessed and documented; review of pre- and post-retraining performance metrics on a held-out evaluation dataset; review of pre- and post-retraining fairness metrics; confirmation that the change is being deployed to staging before production; and documentation of the rollback procedure if performance is unacceptable in production.

Diagram 6.3: AI Change Control Workflow

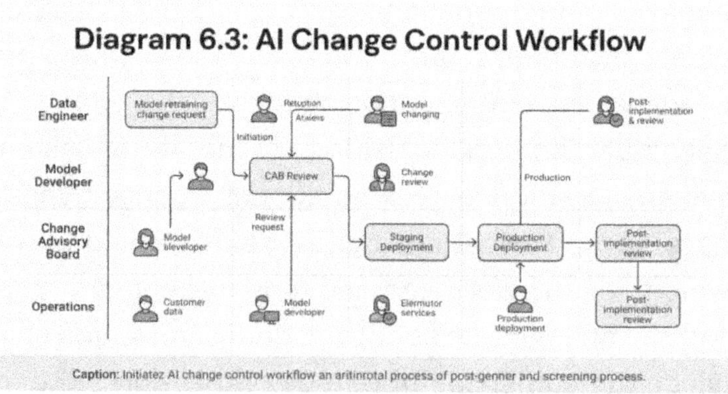

Caption: Initiatez AI change control workflow an aritinrotal process of post-genner and screening process.

Data distribution shifts do not arise from deliberate change decisions and cannot be fully anticipated or prevented. The governance response to distribution shifts is monitoring and response rather than pre-change review. The monitoring program should track statistical properties of model input feature distributions, missingness rates, and value ranges, and alert when these properties diverge significantly from the training distribution. When an alert is triggered, the response procedure should assess whether the divergence is sufficient to invalidate the model's performance guarantees, and, if so, suspend the model or apply mitigating measures (such as increased human-review rates) until the model can be retrained on data representative of the current input distribution.

The change advisory board's role in AI change review requires expanded expertise relative to traditional change management. The board needs at least one member who understands the technical characteristics of AI systems well enough to assess whether a proposed retraining or update carries elevated risk. In organizations where that expertise is not represented in the current CAB membership, the AI governance function should designate a technical reviewer to participate in the CAB review of AI changes as either a standing member or a subject-matter expert resource. The board should also define an emergency change pathway for situations where a model must be suspended or rolled back rapidly; waiting for the next scheduled CAB meeting is not an acceptable response to a live model performance failure.

6.6 Configuration Management

Configuration management for AI systems extends the scope of the configuration management database to include AI-specific configuration items that traditional CMDB definitions do not capture. The configuration items that must be tracked for each AI system include: model artifacts (the trained model files, including version identifiers and checksums); training dataset references (identifiers pointing to the specific dataset versions used to train each model version); evaluation dataset references (the held-out datasets used to validate model performance); hyperparameter configurations (the settings used during training that define the model's architecture and learning behavior); feature engineering specifications (the transformations applied to raw data before it enters the model); model serving configurations (the infrastructure and settings governing how the model is deployed and accessed); and integration point registries (records of all systems that consume the model's outputs).

The most critical property of AI configuration management is the ability to reproduce a prior model state exactly. When an incident occurs, whether a performance degradation, a fairness concern, or a regulatory inquiry, the investigation typically requires answering the question: what was the model doing at a specific point in time, and why? Answering that question requires that the model artifacts, the training data, and the configuration settings from that time be all preserved and linked. Organizations that maintain this traceability can conduct investigations in hours. Organizations that cannot reproduce prior model states may never definitively answer root-cause questions and will face regulatory challenges when they cannot provide a complete account of a system's behavior during the incident period.

Diagram 6.4: AI Configuration Management Database Schema

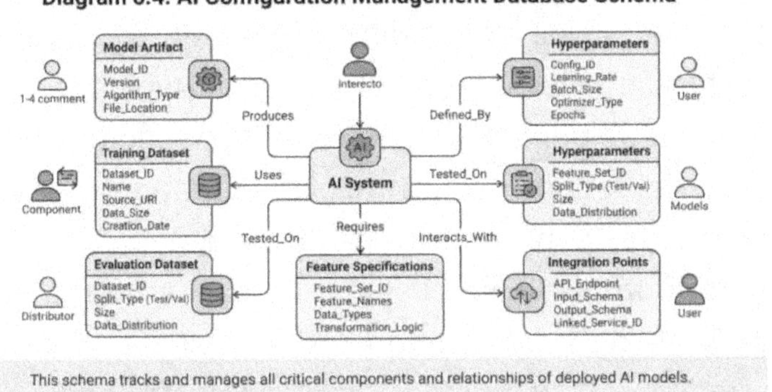

This schema tracks and manages all critical components and relationships of deployed AI models.

The tooling for AI configuration management has matured significantly in recent years. MLflow, DVC (Data Version Control), and similar model lifecycle management tools provide purpose-built functionality for tracking model artifacts, datasets, parameters, and performance metrics across training runs. These tools can be integrated with the organization's existing CMDB or serve as specialized configuration management stores for AI artifacts, feeding metadata summaries into the enterprise CMDB. The choice of tooling is secondary to the requirement: every model in production must have a complete, current, auditable configuration record that can be produced on demand.

Configuration item ownership is as important as configuration item definition. Each CI type in the AI configuration management schema should have a named owner who is accountable for its currency and accuracy. The model development team typically owns the model artifact CI. The data engineering team typically owns the training dataset CI. The system architecture or integration function typically owns the integration point registry. When CI ownership is unclear, records become stale: the configuration management database is only as current as the discipline of its contributors. Establishing clear ownership and embedding configuration record updates into existing operational workflows, making them part of the deployment process rather than a separate administrative task, is the most reliable way to maintain currency.

6.7 Vendor Oversight

The majority of AI capability deployed in enterprise environments comes from vendors: cloud platform AI services, specialized AI software vendors, AI-embedded SaaS applications, and consulting partners who build or operate AI systems on behalf of client organizations. In each of these arrangements, the organization retains legal and regulatory accountability for the outcomes of AI systems it deploys, regardless of whether it built or operates the underlying model. The CFPB's guidance on algorithmic credit scoring makes this explicit: a creditor cannot cite a vendor's black-box model as a defense against fair lending obligations. The organization is responsible, and vendor contracts and oversight programs must reflect that accountability.

Effective vendor oversight for AI begins at the procurement stage with an AI-specific supplier assessment. The assessment should evaluate: the vendor's own AI governance program and whether it meets the organization's standards; the vendor's documentation practices, whether they can provide model cards, training data documentation, and performance evaluation reports. The vendor's incident-response commitments state what happens if the model degrades or produces harmful output. Will the vendor's cooperation with audits allow the organization or its auditors to assess the model's behavior? The vendor's update and change notification practices. How much notice will the organization receive before model changes that might affect compliance?

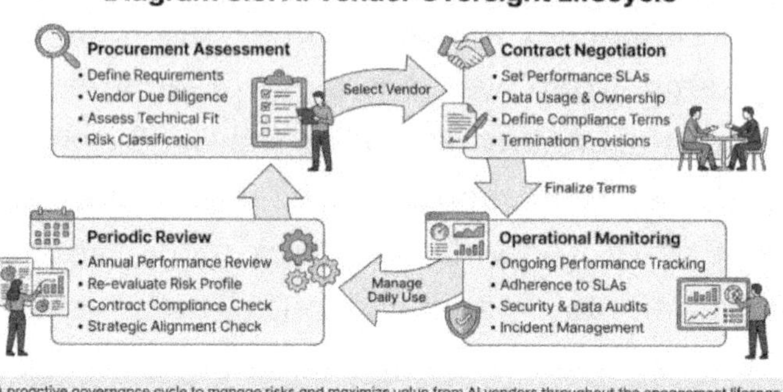

A proactive governance cycle to manage risks and maximize value from AI vendors throughout the engagement lifespan.

Vendor contracts for AI services should contain specific provisions that are not standard in traditional software contracts. These include: model performance SLAs that define acceptable performance thresholds and remedies for degradation; transparency obligations requiring the vendor to provide sufficient documentation for the organization to assess bias, explain decisions, and satisfy regulatory inquiries; change notification requirements with defined advance notice periods for model updates; audit rights allowing the organization or its designated auditors to assess model behavior and governance practices; data processing agreements that address the use of the organization's data in model training; and liability allocation provisions that clarify responsibility for harms arising from model outputs. Legal counsel should review AI vendor contracts specifically for these provisions standard software contract templates do not address them.

Ongoing vendor oversight should be built into the vendor management cadence rather than triggered only by incidents. A defined periodic review annually at minimum, semi-annually for high-risk AI vendors should assess whether the vendor's governance practices remain adequate, whether the model's performance against the contractual SLAs is being maintained, whether any changes to the vendor's business (ownership changes, key personnel departures, product discontinuations) affect the organization's risk posture, and whether recent regulatory developments have changed the compliance requirements the vendor is expected to support. This review should produce a documented finding with a risk rating and, where necessary, a remediation plan negotiated with the vendor.

6.8 Avoiding Duplication

One of the most persistent failure modes in AI governance programs is the creation of governance processes that duplicate existing IT processes without integrating them. When this happens, the same system change may need to go through both the AI governance review process and the IT change advisory board. The same vendor may be assessed by both the AI procurement process and the standard vendor management process. The same risk may appear in both the AI risk register and the enterprise risk register. The result is administrative overhead that burdens the people doing the work, inconsistency when parallel processes reach different conclusions, and erosion of organizational support for governance requirements perceived as redundant.

The solution is deliberate process mapping before governance requirements are finalized. For each AI governance requirement, the design question should be: does an existing process already cover this, does an existing process need to be extended to cover this, or is a new process genuinely needed? The answer is almost always the first or second option, not the third. When an AI governance requirement can be addressed by extending an existing process, extension is strongly preferred because it leverages existing organizational infrastructure, maintains a single authoritative process rather than competing alternatives, and reduces the net governance burden on operational teams.

In practice, avoiding duplication requires active coordination between the AI governance program and the functions that own existing IT processes: the change management function, the vendor management function, the enterprise risk management function, the information security function, and the compliance function. Each of these functions has a legitimate interest in AI governance requirements that fall within its domain. The coordination challenge is to ensure that AI-specific requirements are incorporated into their processes rather than handled separately, and that a clearly designated owner addresses AI-specific requirements without an existing process home. The governance committee is the natural venue for resolving ownership disputes and approving process extensions.

Process integration audits, periodic reviews that assess whether AI governance requirements are being executed within the processes they are assigned to, rather than falling through the gaps, are a practical tool for maintaining integration discipline over time. It is common for process integration to succeed at the design stage and then decay as the processes evolve: a change advisory board that incorporates AI retraining review when the integration is implemented may gradually revert to its default behavior if membership turns over. The AI-specific requirements are not embedded in the process documentation and training materials. The integration audit detects this decay before it becomes a governance gap.

6.9 Operational Discipline

Operational discipline is the quality that distinguishes governance programs that sustain themselves over time from those that degrade after the initial implementation effort. It means that governance requirements are executed consistently, not just when there is visible management attention. It means that documentation is completed contemporaneously, not reconstructed before audits. It means that monitoring alerts trigger responses, not acknowledgments. It means that governance cadences occur on schedule regardless of competing priorities. Operational discipline is not a personality trait; it is an outcome of process design. Well-designed processes with clear ownership, defined timelines, and consequence-bearing accountability tend to produce operational discipline. Poorly designed processes, ambiguous ownership, aspirational timelines, and no consequence for non-execution tend to produce inconsistent execution regardless of the quality of the people involved.

Building operational discipline into an AI governance program requires four design choices. First, each governance activity should have a single named owner, not a team or a committee. Teams and committees make decisions; individuals execute tasks. When execution responsibility is diffuse, execution is unreliable. Second, governance activities should be calendar-driven rather than event-driven. The scheduled risk register review will take place on the first Tuesday of each quarter. A risk register review triggered by a management request will happen only when someone remembers to request it. Third, governance outputs should be visible to management and the governance committee through routine reporting, not only when problems arise. Visibility creates accountability. Fourth, non-execution of governance requirements should have consequences at a minimum, a formal escalation to the governance committee with a documented response and remediation plan.

Diagram 6.6: Governance Cadence Calendar

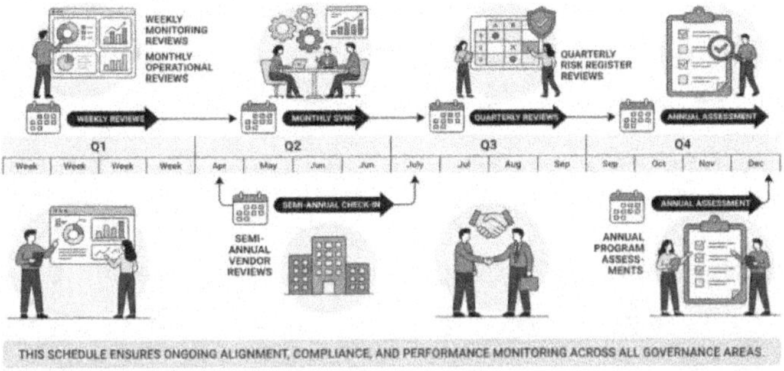

Metrics are the instrument of operational discipline. A governance program that does not track metrics cannot demonstrate its effectiveness and cannot identify where it is underperforming. The core metrics for an AI governance program include: AI system inventory coverage (percentage of production AI systems with current documentation); risk assessment currency (percentage of systems with a risk assessment completed within the defined review cycle); training completion rates by role; monitoring alert response times; control testing coverage and findings rates; and incident response timeliness. These metrics should be reported to the governance committee on a defined schedule and trended over time. Governance maturity is evident in metric trends, including improved coverage, faster response times, and fewer findings, as much as in individual governance artifacts.

The governance program itself should be subject to a formal annual assessment. The assessment should evaluate whether the program's design and governance requirements are appropriate for the current risk profile. Is its operational execution meeting the requirements, and are its outcomes the program's controls actually reducing risk? The annual assessment should produce a written report to the governance committee, with findings prioritized by impact and a remediation plan for each finding. Programs that conduct rigorous annual assessments improve continuously. Programs that do not assess themselves systematically stagnate at whatever maturity level they reached at the end of the initial implementation effort, regardless of how much time passes.

6.10   Manager's Checklist — IT Governance Improvement

To align AI governance with existing IT governance processes, complete the following steps. Conduct a process gap analysis for each major AI governance requirement: map it to the existing IT process that covers the closest analog and document the gap between what the existing process requires and what AI-specific governance needs. Extend the change advisory board charter and intake form to include model retraining events as a change type requiring formal review; train CAB members on the additional review criteria for AI changes. Update the configuration management database schema or implement a specialized ML lifecycle management tool to track model artifacts, datasets, and configurations as configuration items; assign a named owner for each CI type.

Incorporate AI-specific assessment criteria into the vendor management program's supplier assessment template and contract review checklist; ensure legal counsel reviews AI vendor contracts for performance SLAs, transparency obligations, and audit rights. Establish a governance metrics dashboard that reports inventory coverage, risk assessment currency, training completion, monitoring alert response times, and control testing coverage every month. Schedule governance cadence meetings, operational, risk review, vendor review, program assessment, and protect them from displacement by operational priorities. Conduct a process integration audit within 12 months of implementing each AI governance requirement to verify that integration with existing IT processes is being maintained. Review the AI governance process map annually and identify where process duplication has emerged; consolidate into existing IT processes wherever possible.

Diagram 6.7: IT Governance Improvement Roadmap

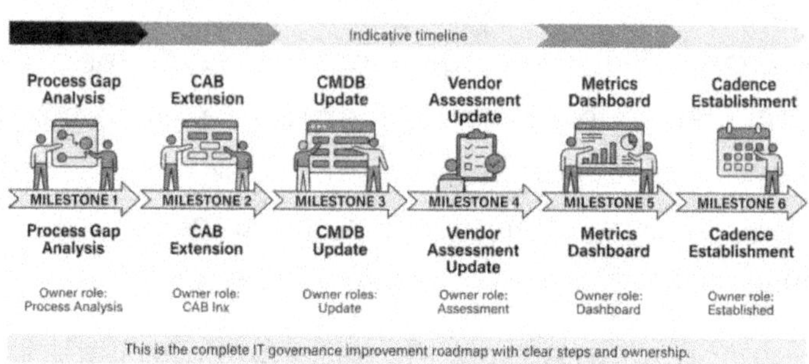

6.11 Key Takeaways

AI governance is most durable when it is integrated into existing IT governance processes rather than maintained as a parallel program. COBIT and ITIL provide the structural vocabulary and process models needed to connect AI-specific requirements to established organizational capabilities. Change control must be extended to cover model retraining events and must include a monitoring-based response mechanism for data distribution shifts. Configuration management must track model artifacts, datasets, and configurations with enough fidelity to support incident investigation and regulatory inquiry response. Vendor oversight must treat AI supplier contracts as a distinct category, requiring specific transparency, audit, and change-notification provisions. Avoiding process duplication is a design requirement enforced through deliberate process mapping and periodic integration audits. Operational discipline, consistent execution, metrics tracking, visible accountability, and annual self-assessment are the qualities that distinguish governance programs that sustain themselves from those that degrade between audit cycles. The goal is not a governance program that looks good on the day of the inspection. It is a governance program that works on an ordinary Tuesday.

7 IT Policy Framework

A regional healthcare network had deployed four distinct AI systems across clinical operations, billing, supply chain, and patient communications over the span of eighteen months. Each project team drafted its own usage rules during the pilot phase, and two teams had written no rules at all. When a billing model began producing erroneous denial codes at scale, the incident response team discovered that no one had documented data handling requirements, approval authorities, or escalation contacts for that system. The post-incident review revealed a more fundamental problem: the organization had acquired AI capabilities faster than it had built the policy infrastructure to govern them. The lesson was not that governance was missing; the organization had a general IT policy library, but nothing in that library addressed the specific decisions, boundaries, and behaviors that AI systems require.

The fallout from that incident extended beyond the immediate technical fix. The organization faced regulatory inquiries because it could not demonstrate consistent data handling practices. Vendor contracts had to be renegotiated because no standard terms had been agreed upon during procurement. Employee confusion about which AI tools were permitted led to a wave of informal shadow AI adoption that IT discovered only through network monitoring. Within six months, what had started as a model performance problem had become a program-level governance crisis. That sequence of events, from uncontrolled adoption to regulatory exposure to cultural breakdown, is not unusual. It is, in fact, the default outcome when AI policy lags AI adoption.

7.1 Why Policy Architecture Matters

Policy is the mechanism by which governance decisions become operational expectations. Without policy, governance committees produce principles that no one is obligated to follow. With a poorly structured policy, managers must interpret conflicting rules or fill gaps through improvised judgment, and auditors cannot verify compliance because there are no documented standards to test against. A coherent AI policy architecture converts the organization's risk appetite and ethical commitments into enforceable, auditable requirements at the workflow level.

The challenge with AI, specifically, is scope. AI governance cuts across domains that have historically been governed separately: IT security, data privacy, vendor management, human resources, legal, and operational risk. A single unified AI policy is rarely workable because each domain has its own vocabulary, audience, and enforcement mechanism. Combining security controls, data classification requirements, model development standards, and acceptable use rules into a single document produces a document that is simultaneously too long for front-line employees to use and too superficial for technical practitioners to implement.

The solution is a modular policy framework: a set of distinct policy documents, each addressing a functional domain, all anchored to a common governance charter that establishes shared definitions, accountabilities, and conflict-resolution rules. Each module is short enough to read in a single sitting, specific enough to be actionable, and owned by the functional leader best positioned to maintain it. The charter provides the architecture that keeps the modules coherent as they evolve independently.

Managers who build a modular policy framework gain three specific advantages. First, accountability is clear: each policy module names an owner and a defined population of employees who must comply, eliminating the ambiguity that occurs when a single document attempts to govern everyone. Second, gaps are visible: when a new AI use case does not map to any existing policy module, that gap triggers a policy development activity rather than being silently ignored, allowing the risk to accumulate. Third, audits are tractable: an external reviewer can examine a single policy document and trace its requirements to a set of controls and supporting evidence without parsing an entire governance library.

Diagram 7.1: Modular Policy Architecture

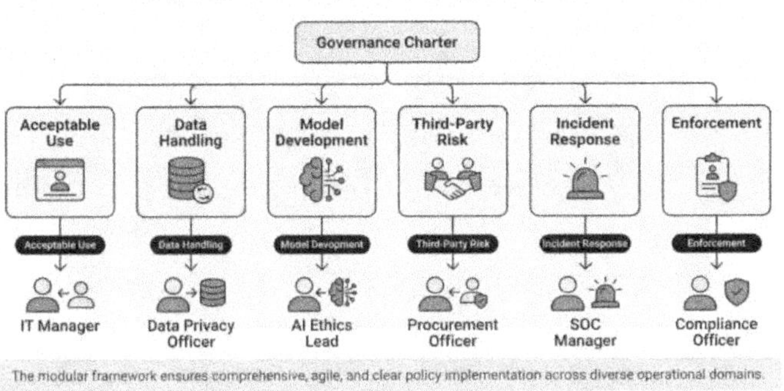

The modular framework ensures comprehensive, agile, and clear policy implementation across diverse operational domains.

7.2 Policy Architecture

A mature AI policy architecture has three tiers. The top tier is a governance charter or policy master document that establishes the scope, principles, decision rights, and relationships among individual policy modules. The middle tier consists of the functional policy modules themselves. The bottom tier contains supporting standards, procedures, and guidelines that provide technical implementation detail without carrying full policy weight. Understanding where each document belongs in this hierarchy prevents the common error of embedding implementation details in high-level policy. This mistake requires frequent policy revisions whenever a technical approach changes.

The charter tier does not contain operational requirements. Its purpose is to state why the organization governs AI, who holds ultimate accountability, and how conflicts between modules are resolved. It defines key terms that appear across multiple modules so that each module does not maintain its own inconsistent glossary. The charter also specifies the review cadence for the policy framework as a whole and the authority required to add, retire, or substantially revise a module. Keeping the charter thin and stable, one to three pages, reduces the frequency with which it must be revised whenever a module is updated.

Policy modules occupy the middle tier and are where most managerial attention should be directed. Each module should answer five questions: what behaviors or activities this policy governs, who must comply, what minimum required practices are, how exceptions are requested and approved, and how compliance is measured. Modules that cannot answer all five questions are incomplete and will lead to enforcement issues. A common incompleteness is a module that specifies required behaviors but never defines who has the authority to approve exceptions. The result is that exceptions are granted informally by whoever is asked, with no tracking or consistency.

The standards tier provides technical specificity. A data handling policy, for example, states that AI training data must be classified and protected according to its sensitivity level. The corresponding data classification standard then defines the sensitivity levels, the protection controls that apply to each, and how classification decisions are made. Separating policy from standard allows technical details to be updated as encryption algorithms or classification categories evolve without requiring a formal policy revision cycle involving legal review and executive sign-off. The principle is that policy tells you what must be true; standards tell you how to make it true.

7.2.1 Policy Ownership and Maintenance

Every policy document requires a named owner who is not the person who wrote the policy but the functional leader responsible for its ongoing accuracy, compliance, and revision. For most AI policy modules, ownership sits with a combination of IT leadership and the relevant business function. The CISO typically owns security-related modules. The Chief Privacy Officer or General Counsel owns the data handling and incident response modules. The CIO or AI program director owns the model development and lifecycle modules. The Chief Procurement Officer or CISO owns the third-party risk module. Shared ownership should be defined carefully: when two leaders share ownership of a document, the charter must specify who has final say when they disagree.

Policy documents should include an explicit review cycle, typically annual, with provisions for off-cycle reviews triggered by regulatory changes, significant incidents, or material changes in the organization's AI portfolio. A specific individual, typically the policy owner, should be responsible for initiating the review and tracking it to completion. A policy that has not been reviewed within its scheduled cycle should be flagged as potentially stale in the governance committee's status reporting and treated with caution during audits. Auditors increasingly ask not just whether a policy exists but when it was last reviewed and whether the review was documented.

Approval authority for policy documents must be matched to the significance of the policy's requirements. A policy that restricts employee behavior or imposes significant compliance obligations on business teams requires a higher level of approval, typically a governance committee vote or executive sign-off, than a technical standard that specifies implementation details within boundaries already established by policy. Establishing these approval tiers in the charter prevents the ambiguity that occurs when managers try to determine whether an update requires a formal vote or can be handled administratively.

Diagram 7.2: Policy Ownership and Maintenance Cycle

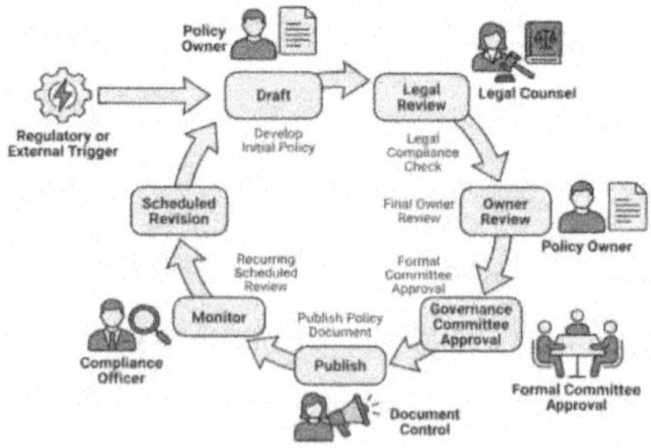

This diagram illustrates the standardized lifecycle for creating, reviewing, approving, publishing, and maintaining organizational policies.

7.3 Acceptable Use Policy

The acceptable use policy defines which AI tools employees may use, under what conditions, and for what purposes. It is typically the most visible AI policy because it affects every employee who interacts with AI-enabled systems or consumer-grade AI tools. Its primary function is to prevent uncontrolled adoption of shadow AI while providing a legitimate, supported path for employees who want to use AI productively. An AUP that is purely prohibitive, one that bans AI tool use without providing approved alternatives, will be ignored because employees will find ways to use tools that make their work easier, regardless of the policy.

The minimum required elements of an effective AUP include: a definition of what constitutes an AI tool for purposes of the policy, because the boundary between AI and non-AI software is increasingly difficult to draw; a list or category description of approved tools; a statement of prohibited uses including prohibited data types that may not be entered into AI systems; a clear rule about employee-sourced or consumer AI tools; and explicit guidance on how employees request approval for tools not yet on the approved list. The definition element is frequently underinvested. A policy that defines AI tools as "systems using machine learning or large language models" will leave employees uncertain about whether AI-enabled features embedded in productivity software they already use are covered.

Prohibited use categories deserve particular care in drafting. The most operationally significant categories include entering regulated personal data health records, financial information, Social Security numbers, children's data into consumer AI systems that process inputs for model training; using AI to make or influence employment decisions such as screening, promotion, or termination without HR and legal review; using AI to generate external communications or legal documents without human review and authorization; and using AI-generated content in regulated disclosures, filings, or certifications without verification. The policy should also explicitly state that existing ethics and conduct rules apply to AI-assisted work: using AI to produce discriminatory content, to deceive customers, or to circumvent internal controls is a policy violation, regardless of whether the AI tool itself is on the approved list.

7.3.1 Approved Tool Registry

The AUP functions most effectively when paired with a maintained registry of approved AI tools. The registry lists each tool, its approved use cases, any conditions or restrictions that constrain its use within the organization, the date of the most recent security and privacy review, and the owner responsible for that tool's continued approval status. Employees can consult the registry before adopting a new tool, reducing the volume of ad-hoc approval requests while making the approval process transparent and predictable.

Registry maintenance is itself a governance function that must be resourced. When a vendor releases a significant product update, the security and privacy properties of that tool may change, including new data-processing capabilities, revised terms of service, and altered model-training practices. The registry owner is responsible for monitoring vendor communications, triggering re-review when material changes occur, and removing tools from the approved list when they fail re-review. The AUP should specify that continued use of a tool after its approval has lapsed is a policy violation, not merely an administrative inconvenience, even if the tool was previously fully approved.

The registry should include a field indicating the scope of use cases for each approval. A tool approved for drafting internal communications is not automatically approved for generating code, processing customer data, or producing financial analyses. Scope-limited approvals enable the organization to increase AI productivity while maintaining targeted controls on the highest-risk use cases. When employees request scope extensions, those requests should go through the same review process as initial approvals, ensuring the highest-risk uses receive the most scrutiny.

Diagram 7.3: Approved Tool Registry Workflow

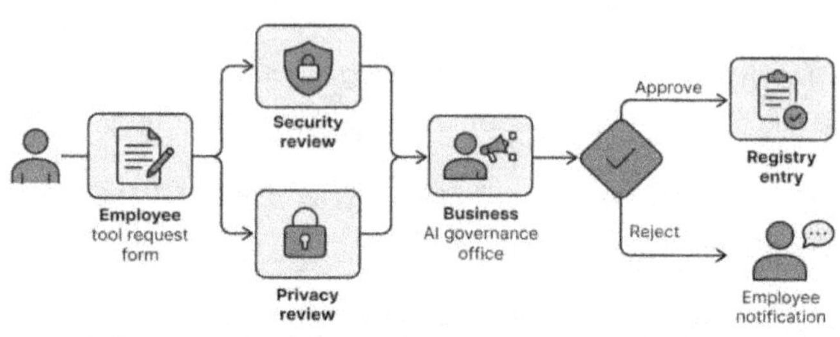

This structured approval process ensures secure and compliant tools for the organization.

7.4 Data Handling Policy

The data handling policy governs how data used to train, test, or operate AI systems is collected, stored, processed, shared, and disposed of. It is functionally distinct from the organization's general data governance policy because AI data handling involves specific practices, such as training data curation, feature engineering, embedding storage, and inference logging, which general data policies do not contemplate. Attempting to govern AI data handling entirely through a general data policy produces gaps that become visible only during incidents or audits.

The policy must address data sourcing with specificity. Not all data sources are equally appropriate for AI use. Publicly sourced data may carry licensing restrictions that prohibit commercial use. Customer data collected for one stated purpose may not be legally permissible for AI training under privacy laws that require purpose limitation. Employee data used to build workforce analytics models raises distinct legal and labor relations considerations. The data handling policy must specify what approval is required before each data source may be used in AI development, what documentation must be created to establish provenance, and who is responsible for making the sourcing determination.

Data quality requirements belong in the data handling policy at the principal level. The policy should not prescribe specific statistical thresholds for completeness or for balance; those should be set in technical standards. Still, it must require that quality be measured, documented, and verified before data enters a production AI system. This requirement obligates building quality checks into development workflows rather than treating data quality as a technical detail that practitioners may address at their discretion. A policy requirement, even at the principal level, creates an audit obligation that organizational behavior must satisfy.

Retention and deletion rules for AI data pose challenges distinct from those in general records management. A model trained on employee performance records may encode statistical patterns that effectively retain information about those individuals in its parameters even after the original records are deleted. This is not a theoretical concern: regulators and courts have increasingly recognized that models trained on personal data may carry privacy obligations that survive deletion of the source data. The data handling policy should acknowledge this complexity and require that model retirement procedures specifically evaluate residual data exposure in model parameters, not just confirm deletion of training files and pipeline artifacts.

7.4.1 Inference and Output Logging

One data-handling question unique to AI systems is how to treat inference logs: records of the inputs submitted to a model and the outputs it produces. Inference logs are valuable for auditing; they enable the reconstruction of past decisions, model debugging, and performance monitoring over time. They are also, in many cases, sensitive personal data when the inputs contain information about employees, customers, or third parties. The data handling policy must specify retention periods for inference logs, access controls that restrict who may review them, and the permitted purposes for which logs may be used beyond operational monitoring.

Output logging raises a related question: when AI systems generate recommendations, predictions, or decisions that employees act on, what obligation exists to retain both the output and a record of its use? Regulatory trends in high-stakes domains such as healthcare and financial services increasingly require not just that automated decisions be logged but also that the human review process, when it occurs, be documented. The policy should specify logging requirements that are calibrated to the risk level of the use case, with more stringent requirements for uses that affect significant interests.

Diagram 7.4: Data Handling Policy Scope and Flow

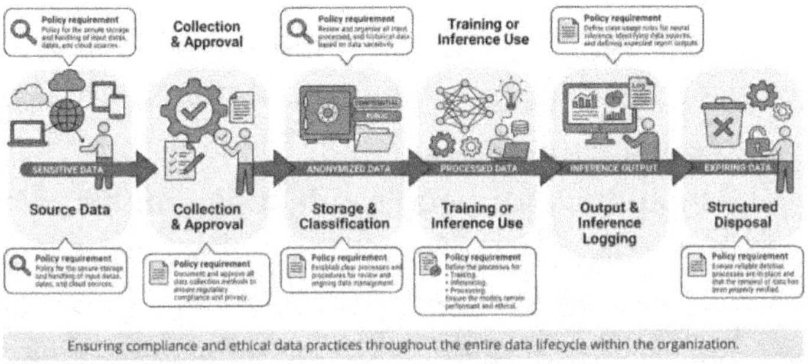

7.5 Model Development Policy

The model development policy establishes the minimum governance requirements for creating any AI model built or significantly customized within the organization. It applies to data science teams, machine learning engineers, and business analysts who build predictive or generative models, regardless of the tools or platforms they use. It does not apply to the use of unmodified pre-built models obtained from vendors, which is addressed under the third-party risk policy. Fine-tuning, embedding proprietary data, or substantially reconfiguring a vendor model typically brings that model within the scope of the model development policy, and the policy should explicitly define this boundary.

At minimum, the policy should require that all model development activities occur within approved development environments that have been provisioned with appropriate access controls and are visible to IT security monitoring; that model cards or equivalent documentation be created for every model that progresses beyond the experimental stage, capturing the model's intended use, known limitations, training data characteristics, and evaluation results; that validation testing including, for people-affecting models, bias and fairness testing be performed and documented before deployment; and that a named individual accept personal accountability for each deployed model's governance compliance.

The policy should clarify the distinction between fine-tuning foundation models and purpose-built model development. Fine-tuning a large language model on proprietary documents, for example, inherits many of the capabilities and limitations of the underlying foundation model. Teams that fine-tune models may not appreciate that they have assumed accountability not only for the fine-tuning layer but also for the system's governance properties, including the base model's data processing practices and potential biases. The policy should specify that governance documentation for fine-tuned models must account for both layers.

Fairness and bias testing requirements in the policy should be stated at a principle level rather than as specific statistical thresholds. The policy should require that fairness considerations be evaluated as part of the validation process for any model whose decisions or recommendations affect people, and that evaluation results be documented and reviewed by the appropriate governance authority before deployment. The definition of a use case that "affects people" should be interpreted broadly: workforce scheduling models, loan risk models, content ranking algorithms, and clinical decision support systems all affect individuals in ways that warrant explicit fairness consideration, even when the connection between model output and individual outcome is mediated by human review.

7.5.1 Experimental Versus Production Models

A common and consequential failure mode in model development governance is the lack of a formal, enforceable boundary between experimental and production use. A team builds a prototype for a limited pilot involving a handful of users. The pilot produces good results, so it expands informally to more users, higher stakes, and broader data access without ever having passed through a formal deployment review. Eventually, the prototype is making production decisions at scale without adequate documentation, monitoring, or accountability structures in place.

The model development policy must define clear, observable criteria to distinguish experimental from production use and specify that transitioning from experimental to production requires a documented release review. Useful criteria for distinguishing experimental from production include: the size and representativeness of the affected population; whether the model's outputs influence decisions with legal, financial, or safety consequences; whether the model operates without continuous human oversight; and whether the development environment uses production data rather than a representative sample. When any of these criteria are met, the experimental label is no longer appropriate, and a release review must be completed before the deployment continues.

Diagram 7.5: Model Development Policy Gate Process

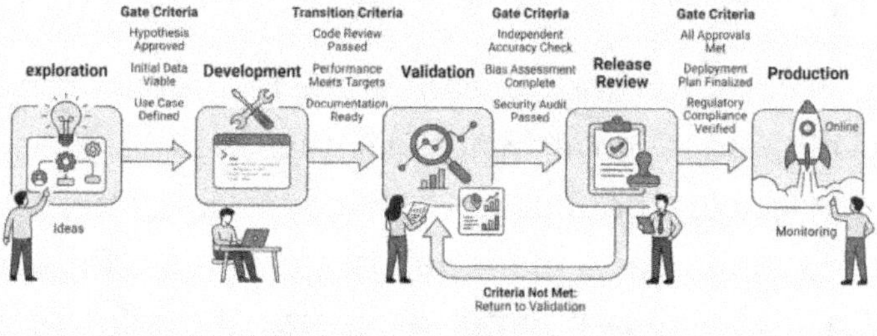

A structured pathway ensuring rigorous checks and approvals throughout the model lifecycle.

7.6 Third-Party Risk Policy

Most organizations use significantly more AI capability through vendors than they build themselves. A third-party AI risk policy governs the procurement, assessment, ongoing monitoring, and offboarding of AI systems and services obtained from external providers. Without this policy, vendor AI enters the organization through procurement processes designed for conventional software: they may verify financial stability, service-level commitments, and basic security hygiene, but they do not examine governance maturity, fairness properties, or data processing practices that distinguish AI systems from conventional software.

The policy must require that AI-specific risk assessment be completed before any AI vendor contract is executed. The assessment should cover, at minimum, the vendor's data processing practices what data the system collects, how it is used, whether it is used to improve vendor models; model accuracy and limitation disclosures, including known performance gaps across demographic groups; incident notification obligations in the vendor's standard terms; the vendor's own governance and ethics program maturity; contractual audit rights that allow the organization to verify vendor compliance with agreed terms; and data deletion obligations upon contract termination. Standard IT vendor assessments rarely cover all of these dimensions, and relying on them can create a false sense of assurance.

Vendor risk tiering is an important design choice for the policy. Not all AI vendors pose equivalent risk, and applying the same intensive assessment to a simple text classification tool as to a high-stakes clinical decision support system creates unnecessary friction without improving governance. The policy should define risk tiers based on factors such as the sensitivity of the data processed, the stakes of the decisions affected, the degree of human oversight in normal operation, and the vendor's reliance on proprietary AI that cannot be easily replaced. Higher-tier vendors receive more intensive initial assessment and more frequent ongoing monitoring.

Ongoing monitoring is as important as initial assessment and is frequently the weaker element of vendor AI governance programs. AI vendor products change frequently, sometimes through automatic updates that alter model behavior, data processing scope, or terms of service without prominent customer notification. The policy should require periodic re-assessment of high-risk AI vendors on a defined schedule at a minimum of annually, more frequently for the highest-risk vendors specify how material changes disclosed by the vendor trigger an immediate re-assessment, and define an accelerated review process for situations where a vendor is subject to a security incident, regulatory action, or significant adverse coverage related to its AI systems.

7.6.1 Contractual Minimum Requirements

The third-party risk policy should specify minimum contractual requirements that must be present in every AI vendor agreement. Legal counsel should review these provisions as AI-specific contract terms evolve in industry practice. At minimum, the required provisions should include: an explicit definition of what data the vendor may access and process, and for what purposes; a prohibition on using customer data to train or improve vendor AI models without the organization's explicit written consent; incident notification timelines that are consistent with the organization's regulatory obligations; audit rights that allow the organization to request documentation of the vendor's data processing and model governance practices; and specific data return and deletion procedures upon contract termination, including a provision addressing any data that may have been incorporated into vendor model parameters during the contract period.

Organizations should maintain a standard AI vendor contract addendum that contains these provisions in pre-negotiated form. The addendum reduces legal review time for routine procurements. It ensures that contract terms are consistent across the vendor portfolio rather than varying based on negotiating leverage and legal attention in individual deals. When vendors refuse to accept standard provisions, the variance should be escalated to an appropriate authority and documented to create an audit trail that demonstrates the organization's attempt to maintain standards even when commercial negotiations result in compromises.

Diagram 7.6: Vendor AI Risk Assessment and Tiering

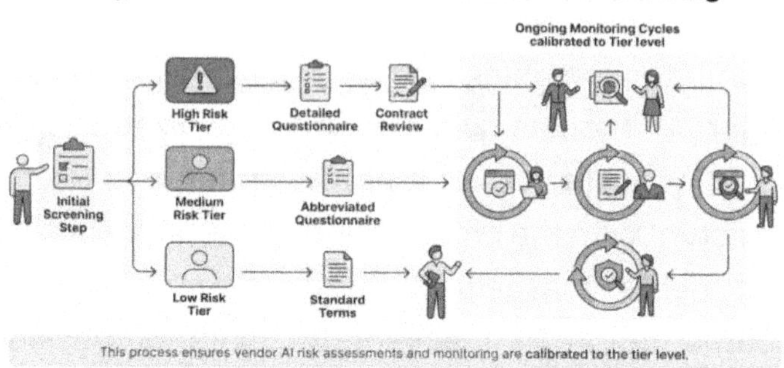

This process ensures vendor AI risk assessments and monitoring are calibrated to the tier level.

7.7 Incident Response Policy

AI incidents differ from conventional IT incidents in ways that require deliberate policy adaptation. A cybersecurity incident typically has a clear triggering event: a breach, a malware infection, or unauthorized access. The onset is usually identifiable with some precision. AI incidents are often gradual: a model's performance degrades slowly as the data distribution shifts, a bias pattern emerges over months of production use, or an automated decision process begins producing systematically incorrect outputs without triggering any alert because the error rate is below any configured monitoring threshold. The incident response policy must account for both acute failures and degradation-type events that may develop over extended periods before becoming visible.

The policy should define with specificity what constitutes a reportable AI incident. Ambiguity in this definition leads to under-reporting, which in turn prevents the organization from maintaining accurate awareness of its AI risk exposure. Suggested categories include: model outputs that cause material harm to customers, employees, or third parties; security incidents affecting AI infrastructure, training data pipelines, or model storage; discovery that a model's performance has fallen below a defined operational threshold on a monitored metric; outputs that appear to reflect discriminatory patterns based on protected characteristics; and incidents where AI-generated content has been published, communicated to external parties, or acted upon in ways that create legal, regulatory, or reputational exposure. For each category, the policy should specify the notification chain, the documentation requirements, and the escalation threshold that triggers executive or board notification.

Recovery from AI incidents often requires decisions and actions beyond the scope of standard IT recovery playbooks. An AI incident may require taking a model permanently offline rather than restoring it from backup, because the issue may be in the model itself rather than in supporting infrastructure. It may require reverting to a prior model version, which itself must have been properly versioned and preserved, or temporarily replacing an automated decision with a manual process while the AI system is under investigation. The policy must specify who has authority to make each of these decisions and under what time constraint, ensuring that high-stakes decisions do not wait for approvals that normal IT recovery procedures were not designed to provide.

The relationship between the AI incident response policy and the organization's broader incident response and business continuity frameworks must be explicitly defined. Many organizations have mature cybersecurity incident response plans and business continuity plans. The AI incident response policy should build on these frameworks rather than creating a parallel structure. It should identify which existing escalation paths and communication templates apply to AI incidents and specify where AI-specific procedures supplement or modify the general framework. This integration prevents the confusion that arises when responders must decide in real time which procedure takes precedence.

7.7.1 Post-Incident Review Requirements

Every reportable AI incident should result in a documented post-incident review that examines not only the technical aspects but also the governance, policy, and organizational factors that contributed to the incident and its severity. The review should specifically ask whether the existing policy was violated, whether the policy was adequate but not followed, or whether the incident reveals a gap in policy coverage that a new or revised module must address. Reviews of this type generate evidence that drives policy improvement over time. They also create a record when compiled into an annual trend analysis of how the organization learns from its AI failures, which is precisely the kind of learning-culture evidence that regulators and auditors find compelling when evaluating the maturity of a governance program.

Diagram 7.7: AI Incident Classification and Response Flow

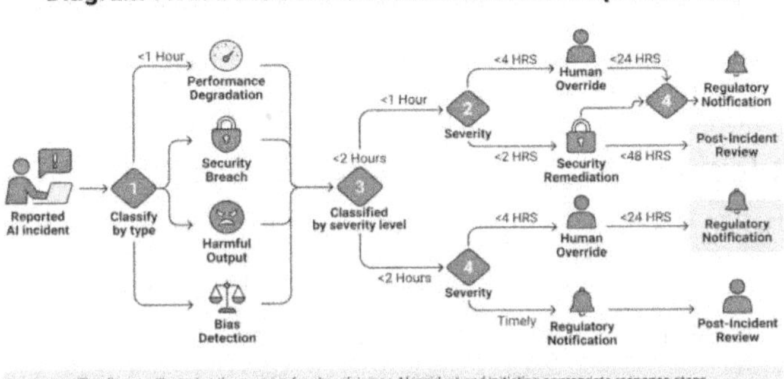

The diagram illustrates the process for classifying an AI incident and initiating appropriate response steps.

7.8 Enforcement and Exceptions

A policy that cannot be enforced is a liability. It creates a documented standard that the organization visibly fails to meet, which is worse for audit and litigation purposes than having no documented standard at all. Building an enforceable AI policy framework requires attention to three interdependent elements: monitoring mechanisms that detect non-compliance before it escalates to material risk, consequence structures that are proportional and consistently applied, and exception processes that provide a legitimate path for justified deviations without creating a permanent loophole culture.

Monitoring for policy compliance in an AI context is technically more complex than in general IT governance.

Employees may use AI tools through personal devices or consumer accounts that are not visible to IT monitoring systems. Data scientists may build models in informal notebook environments that do not pass through a centralized repository or approval workflows. The compliance monitoring approach must acknowledge these realities and prioritize monitoring of the highest-risk activities, use of sensitive data in AI systems, deployment of models without documented review, vendor contracts executed without AI-specific assessment, rather than attempting comprehensive surveillance that is both technically impossible and culturally counterproductive.

Consequence structures must be proportional to the severity of violations and applied consistently across the organization. Policies that prescribe consequences only for the most severe violations and treat minor non-compliance as an inconsequential administrative gap will be treated by employees as aspirational rather than mandatory. The governance framework should include a graduated enforcement model: first-time or minor violations receive documented guidance and a correction requirement; repeated violations or moderate-severity non-compliance receive formal corrective action with defined remediation timelines; and material violations or intentional circumvention receive escalation to HR, legal, or the executive team as appropriate. The critical feature of a graduated model is that every violation is documented, not just the severe ones, creating a record that supports both pattern analysis and escalation decisions.

Exception processes are the pressure valve that prevents well-designed policy from becoming an obstacle to legitimate business activity. Every policy module should define an exception request process that specifies how to submit a request, who reviews it, what justification is required, and what compensating controls must be in place. In contrast, the exception is active and specifies how long it remains valid before requiring renewal. Exceptions should be tracked centrally, not managed informally by individual policy owners, so that the governance committee can identify patterns. A module that consistently generates high exception volume may be poorly calibrated to actual operational needs. That signal should drive policy revision rather than continued approval of individual exceptions.

7.8.1 Metrics for Policy Effectiveness

The policy architecture should include a metrics program that tracks indicators to determine whether the policy is actually governing behavior. Useful metrics include: compliance rate by policy module as measured through periodic surveys and technical monitoring; exception volume and trend by module over time; mean time to remediation for identified violations; percentage of deployed AI systems that have a current, reviewed policy mapping; percentage of AI vendor contracts that include the required minimum provisions; and the number of AI incidents that post-incident reviews attribute wholly or partly to policy gaps or non-compliance. These metrics should be reported to the governance committee at regular intervals and used to drive both policy revision and resource allocation.

Governance committees that receive only summary policy status reports, "all policies are current, compliance is satisfactory," cannot identify where the real risks are accumulating. Metrics that show trends over time, variation across business units, and correlations between policy maturity and incident frequency enable the committee to make informed decisions about where to allocate governance attention and resources. The metrics program is therefore not an administrative overhead but a core tool for directing governance improvement.

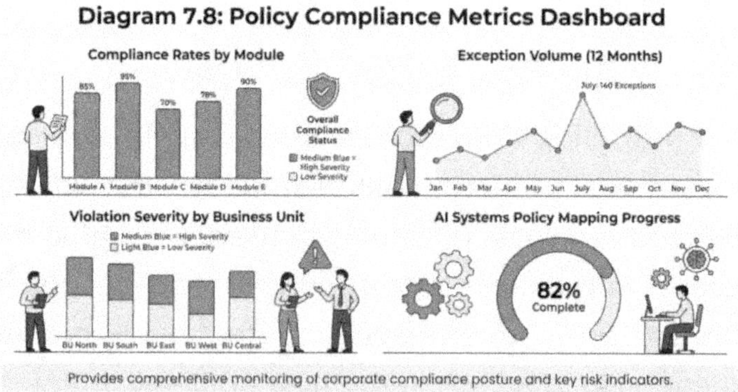

Diagram 7.8: Policy Compliance Metrics Dashboard

Provides comprehensive monitoring of corporate compliance posture and key risk indicators.

7.9 Manager's Policy Framework Checklist

The following behaviors and artifacts mark a manager who has built a functional AI policy framework. Review this list at the start of each annual policy cycle and after any significant AI incident or regulatory development.

A governance charter exists that names the owner and review cycle for every AI policy module and specifies how inter-module conflicts are resolved. Every active AI system in the production environment can be mapped to at least one policy module that governs it, with no systems operating outside documented policy coverage. HR and legal counsel have reviewed the acceptable use policy within the past twelve months, employees have formally acknowledged it, and an approved tool registry is maintained and current. The data handling policy specifically addresses AI training data sourcing requirements, inference and output logging, and data obligations triggered by model retirement. The model development policy defines the criteria for the transition from experimental to production and requires model cards with documented validation results before deployment. Third-party AI vendors have been assessed against AI-specific criteria, not just general IT vendor questionnaires, and vendor contracts contain the required minimum provisions. The incident response policy defines AI-specific incident categories, response authorities, and post-incident review requirements. An exception log is current, all exceptions are time-bounded and include specified compensating controls, and you have reviewed exception volume trends in the past quarter. You can produce a metrics report showing compliance rates by policy module and exception trends upon request at any time.

7.10 Building Policy That Governs in Practice

The test of a policy framework is not whether it looks comprehensive on paper but whether it changes behavior at the point of AI adoption, development, and operation. A policy that is unknown to the people it governs, too complex to follow in the course of normal work, or unenforced in practice, provides the illusion of governance without its substance. Organizations that discover this gap, usually during a regulatory inquiry or material incident, face the dual burden of proving retrospective compliance with standards they cannot demonstrate were ever consistently followed, while simultaneously rebuilding a framework that should have been in place from the start.

Building policy that actually governs requires a sequence of investments: starting with the highest-risk gaps rather than attempting to address every possible scenario at once; writing requirements in plain language that practitioners and front-line employees can follow without legal interpretation; assigning ownership at the managerial level where accountability for outcomes is also held; investing in the monitoring and exception management infrastructure that makes compliance a realistic expectation; and treating every incident and exception pattern as a signal that drives improvement rather than an embarrassment to be minimized. Policy maintenance is not an annual document-refresh exercise. It is a continuous feedback loop between written requirements and observed organizational behavior.

Modular policy architecture scales well. As the organization adds new AI capabilities, such as generative AI tools, autonomous decision systems, and embedded AI in enterprise software, new modules can be added without rewriting existing ones. As regulations evolve, the affected modules can be revised without destabilizing the entire policy library. The governance committee can track coverage, identify gaps, and prioritize development resources based on a clear map of what is and is not currently governed. That visibility, knowing precisely where the rules exist and where they do not, is itself a governance achievement. An organization that can accurately describe the boundaries of its policy framework is in a fundamentally stronger governance position than one that believes its policies are comprehensive when, in fact, they are not.

Diagram 7.9: Policy Framework Maturity Progression

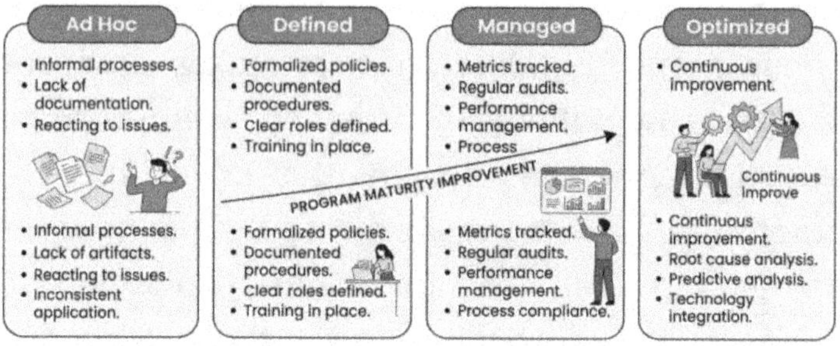

A systematic approach to developing and evolving organizational policy frameworks over time.

8 AI Development Life Cycle

An insurance company's data science team spent nine months building a claims severity model that, by internal evaluation, was among the most accurate the organization had ever deployed. The model's performance on holdout test sets was strong, the development team had carefully documented its architecture, and the business sponsor was enthusiastic about going live. When the model reached the compliance review stage, an ad hoc review was added late in the project following a previous regulatory inquiry, and the reviewers discovered three problems. The training data had not been documented with sufficient provenance to satisfy the company's own data handling policy. The model had never been evaluated for performance disparities across demographic groups. And no one had defined what monitoring would be in place after deployment or who would own the model's ongoing performance. The project was delayed by six months while these gaps were remediated, at high cost and with frustration for the team.

The underlying cause was not negligence. The data scientists were skilled practitioners who had built a technically sound model. The cause was structural: governance requirements had not been integrated into the development workflow. They existed as a separate compliance gate at the end of the project rather than as checkpoints embedded throughout the process. This meant that governance was applied retroactively to decisions that had already been made and were expensive to reverse. Integrating governance into the AI development lifecycle at each stage, as a condition for progressing to the next, is the mechanism that prevents this pattern.

8.1 Why Lifecycle Governance Matters

The AI development lifecycle is a sequence of decisions, each of which shapes the properties of the final system. Training data selection determines what patterns the model can learn and what biases it may encode. Feature engineering choices determine what information the model uses to make predictions and what factors it treats as irrelevant. Model architecture choices determine the tradeoff between accuracy and interpretability. Evaluation metric selection determines what constitutes "good performance" and may obscure poor performance in specific subgroups. Deployment infrastructure choices determine how quickly the model can be updated and monitored, and, if necessary, rolled back.

Governance applied only at the end of the lifecycle cannot meaningfully influence most of these decisions. By the time a model reaches a final compliance review, the training data has been collected, the features have been engineered, the architecture has been selected, the model has been trained and evaluated, and the team has made dozens of consequential choices under the assumption that the approach is sound. A late-stage review that identifies a problem with any of these choices: the training data lacks provenance documentation; the evaluation methodology failed to test for demographic disparities; and the process requires going back to an earlier stage and repeating subsequent work. The cost of late governance is therefore not just the time spent in review, but the rework triggered by findings that an earlier checkpoint would have caught and resolved much more cheaply.

For managers, the practical implication is that lifecycle governance must be designed into the development process as a series of stage-gate requirements rather than appended as a post-development review. Each stage must have defined entry criteria conditions that must be satisfied before work at that stage may begin, and exit criteria conditions that must be satisfied before the team may proceed to the next stage. Meeting these criteria requires documentation, and producing that documentation is how governance evidence accumulates during development rather than being reconstructed after the fact.

Diagram 8.1: AI Development Lifecycle Governance Overview

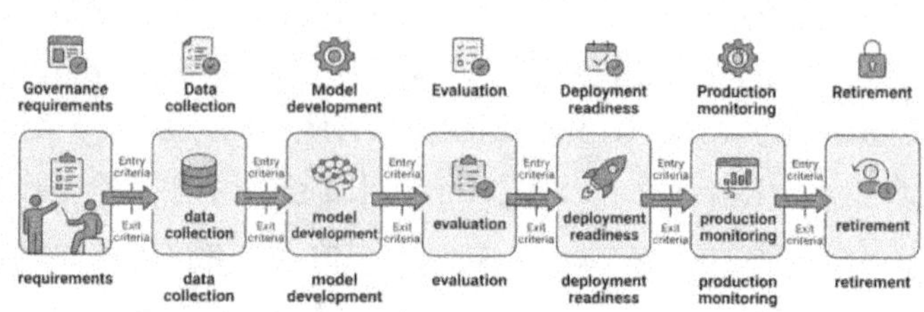

Caption: AI responsibility and compliance are key and remain at the core of all business processes.

8.2 Lifecycle Overview

The AI development lifecycle, as described in this chapter, has seven stages: requirements and scoping, data collection and provenance, model development controls, evaluation and validation, deployment readiness, production monitoring, and retirement. These stages are not purely sequential in practice; data collection often informs initial feature engineering, and evaluation findings frequently require returning to model development for iteration. However, the stage-gate governance model remains valid even in iterative workflows: each time a team prepares to move from one phase to the next, the exit criteria for the current phase must be met.

Different AI development methodologies, such as agile data science, CRISP-DM, and MLOps-based continuous development, use different vocabulary for similar stages and emphasize iteration differently. The governance framework in this chapter is designed to be methodology-neutral: it specifies what must be documented and verified at each decision point, not how the technical work should be organized. Teams using agile sprints can satisfy governance requirements within each sprint cycle. Teams using a more waterfall-structured approach can apply them as formal phase gates. The requirement is that governance checkpoints occur before consequential decisions are finalized, regardless of the development methodology.

One important design principle for lifecycle governance is that the documentation burden should be calibrated to the risk level of the AI system being developed. A model that provides low-stakes internal recommendations, operates under continuous human review, and processes only non-sensitive data warrants a lighter governance documentation requirement than a model that makes autonomous recommendations in healthcare, finance, or employment. The governance framework should define risk tiers and specify the documentation requirements applicable to each tier, so that teams building low-risk systems are not burdened with the same overhead as teams building high-risk systems, and teams building high-risk systems cannot apply the lighter requirements that would be appropriate only for lower-risk work.

Diagram 8.2: Lifecycle Stage-Gate Entry and Exit Criteria

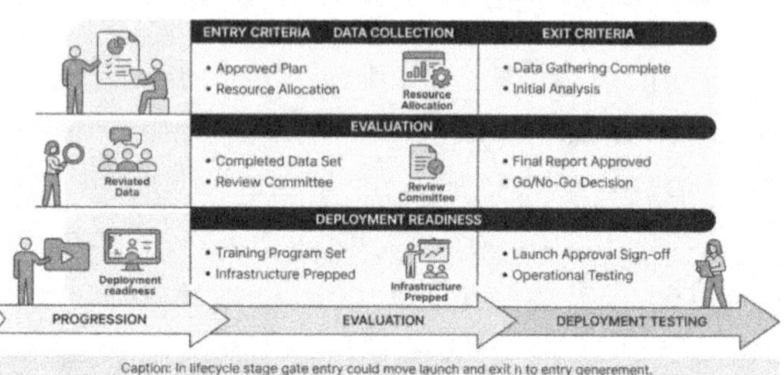

Caption: In lifecycle stage gate entry could move launch and exit h to entry generement.

8.3 Requirements and Scoping

The requirements and scoping stage is where governance has its highest leverage. Decisions made here about the model's intended use, the population it will affect, and the decisions it will influence propagate through every subsequent stage and are extremely expensive to reverse once development is underway. Yet governance requirements at the scoping stage are consistently underinvested, because the stage does not produce a model and therefore does not feel like "real" AI development to practitioners focused on technical output.

The minimum governance documentation required at the scoping stage includes an AI use case charter that specifies the problem the model is intended to solve, the population or assets the model's outputs will affect, the business process the model will be integrated into, the intended decision authority of the model will it make decisions autonomously, provide recommendations for human review, or flag cases for escalation and the criteria that will be used to evaluate whether the model has achieved its intended purpose after deployment. The charter should be approved by both the technical lead and the accountable business owner before data collection begins.

Risk classification must occur at the scoping stage. The risk classification determines which documentation requirements and governance controls apply to the development project. Classification criteria should include the sensitivity of the data the model will process, the significance of the decisions it will influence, the size of the affected population, the degree of human oversight in normal operation, and any applicable regulatory designations for the use case. A model classified as high risk should trigger additional requirements at every subsequent lifecycle stage, including more extensive data-provenance documentation, more rigorous validation testing, and mandatory fairness evaluation, while a low-risk model operates under a lighter but still documented governance process.

8.3.1 Defining Intended Use and Misuse

An often-overlooked element of requirements governance is the explicit documentation of intended use and anticipated misuse. Intended use documentation specifies the contexts in which the model is designed to operate: the population it was trained on, the conditions under which its predictions are expected to be reliable, and the boundaries of appropriate application. Documentation of anticipated misuse specifies how the model could be used beyond its intended scope and the controls that will prevent or detect such use. Both documents should be produced at the scoping stage and maintained throughout the lifecycle, because they form the basis for evaluating whether the deployed model is being used appropriately.

Risk-tiered use-case scoping also helps the organization prevent scope creep after deployment. When a model is later proposed for a new use case, the original scoping documentation serves as an authoritative reference for evaluating whether the new use case falls within the intended design parameters. If it does not, a new scoping and development process must begin rather than extending the existing model beyond its validated boundaries.

Diagram 8.3: AI Use Case Risk Classification Matrix

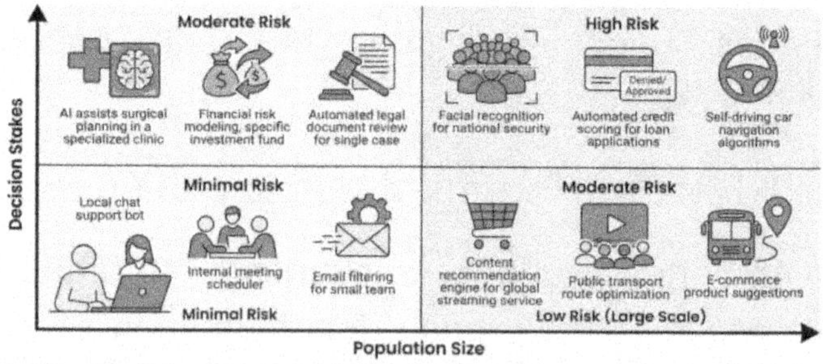

A structured approach to classifying and managing AI risks based on impact severity and scale of affected population.

8.4 Data Collection and Provenance

Training data is not a neutral input. The patterns encoded in training data, including historical biases, representation imbalances, and labeling errors, become features of the model. Data governance at the collection stage is therefore not an administrative formality but a direct influence on the model's eventual behavior. The data collection and provenance stage requires documentation of where the data came from, what transformations were applied to it, its quality characteristics, and the legal and policy constraints on its use in model training.

Provenance documentation should create a complete, auditable record of data lineage from source to training set. For each data source incorporated into the training data, the documentation should specify: the source system or data provider, the date range of the data, the legal basis on which the data is being used for AI training, any transformations or filtering applied, the version of the data pipeline used to produce the training set, and the individual who authorized the use of each source. This documentation is not primarily for the development team's own reference; they generally know where the data came from. It is for auditors, regulators, and post-incident investigators who may need to reconstruct what was in the training data years after the model was built.

Data quality documentation must go beyond summary statistics. Documenting the mean and standard deviation of numerical features indicates to an auditor that the data were analyzed. Still, it does not reveal whether the data is representative of the population the model will encounter in deployment, whether it contains systematic labeling errors, or whether specific demographic groups are underrepresented in ways that will degrade model performance for those groups. Quality documentation at the governance standard should include: representation analysis by relevant subgroups where applicable; known data gaps or collection biases; the labeling methodology for supervised learning tasks, including inter-annotator agreement metrics where human labeling was used; and the outcome of any data quality checks that were performed and failed, with a description of how the failure was resolved.

Data versioning is a governance control, not merely a technical convenience. When a model must be investigated after a production incident, the ability to retrieve the exact training data used to build that model version, not approximately the same data, but the exact data, is essential for diagnosis. Governance requirements should specify that training data sets must be versioned and stored with retention periods matched to the expected operational life of the model, plus an additional period adequate for post-incident investigation. Deleting or overwriting training data while the model it produced remains in production is a governance violation.

8.4.1 Third-Party and Synthetic Data

Training data obtained from external providers requires specific provenance documentation that addresses the vendor's data collection practices, the licensing terms under which the data may be used for AI training, and any restrictions on the types of models or use cases for which the data is licensed. Open-source training data sets warrant the same scrutiny as commercial data products: many widely used open-source data sets have documented biases, data quality issues, or licensing ambiguities that practitioners may not be aware of at the time of use.

Synthetic data generated programmatically to augment real training data or substitute for it when real data is unavailable presents its own governance considerations. Synthetic data may reduce privacy risk compared to using real personal data. Still, it introduces a new question: does the synthetic data adequately represent the real-world distribution the model will encounter? Documentation for synthetic data use should specify how the synthetic data was generated, what real data it was derived from or intended to represent, and what validation was performed to verify that the synthetic data is an adequate substitute for the real data it replaces.

Diagram 8.4: Data Provenance Documentation Framework

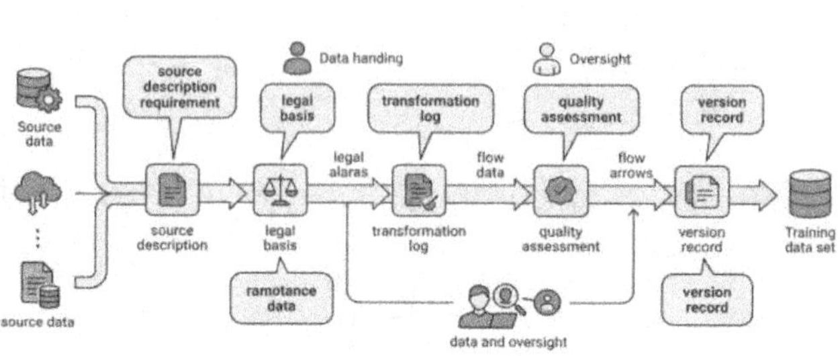

A robust documentation framework is critical for establishing trust and accountability in data provenance for machine learning models.

8.5 Model Development Controls

The model development stage covers the activities between finalized training data and a candidate model ready for evaluation. Governance controls at this stage focus on the integrity of the development environment, code and experiment management, and documentation of key development decisions. The goal is to ensure that the development process is reproducible, so that another engineer could reconstruct the model from the documented inputs and processes, and that significant choices made during development are documented in ways that support subsequent review.

Development environment controls specify that model development must occur in approved, managed environments rather than on personal computers, unmanaged cloud instances, or other computing resources outside the organization's security-monitoring and access-control perimeter. The approved environments should be provisioned with version control systems to capture code changes, experiment-tracking tools to record hyperparameter choices and intermediate results, and access controls to restrict who may modify training pipelines. The rationale for this control is not primarily security. However, security is a consideration, but reproducibility and accountability: a model development process that cannot be reconstructed because it occurred in an unmanaged environment cannot be audited.

Experiment tracking is both a governance requirement and a development best practice. When a team runs dozens of experiments exploring different model architectures, feature combinations, and hyperparameter settings before arriving at the final model, the choices that were not made are as relevant to governance as the choices that were. If the team evaluated a simpler, more interpretable model architecture and chose a more complex one instead, the governance documentation should record that tradeoff and the reasoning behind it. Regulators and auditors in high-stakes domains increasingly ask whether the organization considered less complex or more interpretable alternatives and why they were rejected.

Model cards must be initiated during the development stage, not deferred to the deployment preparation stage. A model card is a structured document that describes the model's purpose, architecture, training data characteristics, performance metrics across relevant subgroups, known limitations, and recommended use cases. Starting the model card during development ensures that the development team, who have direct knowledge of the training process, the architectural choices, and the observed performance characteristics, contribute to the document while that knowledge is current. Model cards completed retrospectively by deployment teams who were not involved in development are invariably less accurate and less useful.

8.5.1 Interpretability and Documentation Standards

The documentation standard for model architecture choices should require the development team to document not just which model was built but also why that architectural approach was chosen over alternatives. For models in high-risk tiers, this documentation should specifically address interpretability: whether a more interpretable model was considered, whether it was evaluated and found inadequate, and if so, why. The principle is that the burden of justification increases with the stakes of the model's decisions. An autonomous decision model in a regulated domain that uses a black-box deep learning architecture should include a documented justification for why interpretability requirements could not be met with a simpler approach.

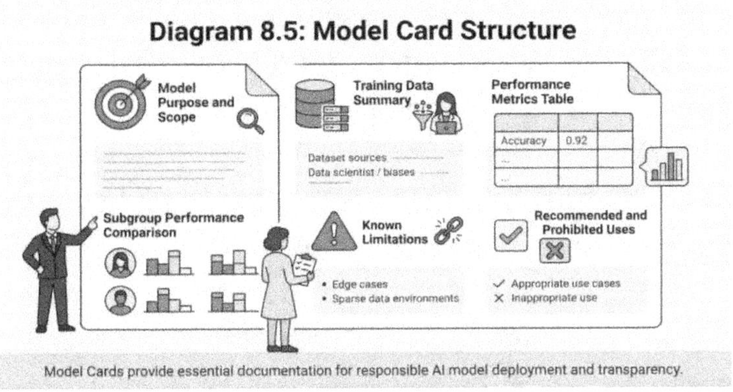

Model Cards provide essential documentation for responsible AI model deployment and transparency.

8.6 Evaluation and Validation

The evaluation and validation stage is where governance requirements are most often contested between data science teams and governance reviewers. Data scientists evaluate models against technical performance metrics such as accuracy, F1 score, and AUC. They may regard additional validation requirements as governance overhead that delays deployment of a model whose performance is already demonstrably strong. Governance reviewers, by contrast, focus on evaluation dimensions that technical metrics do not capture: whether the model performs consistently across demographic subgroups, whether it is robust to the data distribution in the production environment, and whether its performance can be explained to non-technical stakeholders who must make decisions based on its outputs.

The exit criteria for the evaluation stage must be defined before the stage begins. Defining success criteria after evaluation results are known, the practice of adjusting thresholds to match observed results is a fundamental governance failure that produces models with false assurance of adequacy. The AI use case charter prepared at the scoping stage should specify the performance criteria the model must meet. If those criteria are revised during development based on technical findings, the revision must be documented and approved by the business owner, not made unilaterally by the development team.

Fairness evaluation is a required component of the validation process for any model in a medium- or high-risk tier. The form of fairness evaluation appropriate to a given use case depends on the nature of the decision and the relevant protected characteristics. Statistical fairness criteria, demographic parity, equalized odds, and calibration across subgroups may conflict with each other and with overall accuracy, and the governance documentation should reflect a deliberate choice among these criteria rather than the application of a single default metric. The choice should be made by the accountable business owner in consultation with legal counsel and HR, where applicable, not by the technical team alone.

Robustness testing evaluates whether the model performs reliably under conditions that differ from the training distribution. Production environments rarely match training data exactly. Data distributions shift over time, edge cases arise that were not represented in training, and adversarial inputs may be crafted by users seeking favorable model outputs. Robustness testing should include evaluation on out-of-time samples, where available; testing on demographic or geographic subgroups that may be underrepresented in the training data; and, for models exposed to end-user inputs, adversarial testing to identify inputs designed to produce incorrect or harmful outputs.

8.6.1 Independent Validation for High-Risk Models

Models classified as the highest-risk tier should undergo an independent validation review before deployment. Independent validation means evaluation by individuals who were not involved in the development process, either an internal team with no prior involvement or an external third party. The value of independence is that it eliminates the confirmation bias that affects self-evaluation: development teams who believe their model is good will find evidence that supports that belief. At the same time, independent reviewers are more likely to identify weaknesses that the development team has accepted or overlooked. Independent validation reports should be retained as governance evidence and reviewed by the governance committee before deployment approval is granted.

Diagram 8.6: Evaluation and Validation Checklist

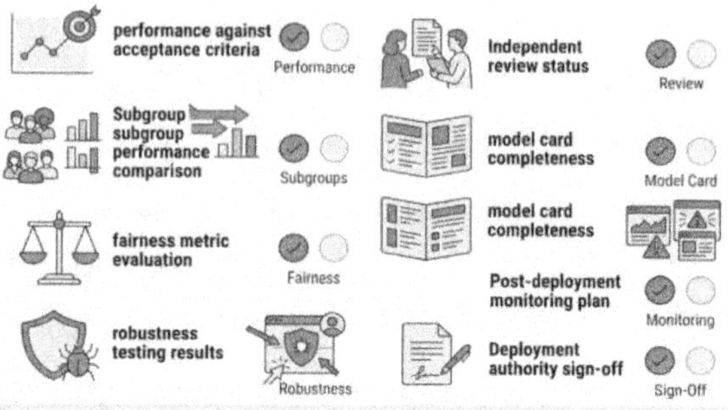

This checklist includes these measures and evaluation list products for integrated deployment and evaluative systems.

8.7 Deployment Readiness

Deployment readiness is the final stage before a model enters production. Its purpose is not to re-evaluate the model that work should have been completed at the evaluation stage, but to verify that the production environment and the organizational context are ready to operate the model safely. Deployment readiness checks are frequently omitted or treated perfunctorily because the model has already passed technical evaluation, and the business sponsor is eager to go live. This creates a category of production incidents in which the model performs as designed, but the surrounding infrastructure, monitoring, or organizational processes are inadequate to support it safely.

The deployment readiness checklist should cover at minimum: production infrastructure review confirming that the serving environment, API, and integration points have been security-reviewed and load-tested; monitoring configuration confirming that performance metrics, data drift indicators, and error rates are being logged to an accessible monitoring system with alert thresholds configured; human oversight process documentation confirming that the process by which model outputs are reviewed, acted upon, or overridden by humans is defined, communicated to the relevant staff, and operational; rollback capability confirmation that the prior production system or model version can be restored within a defined time window if the new deployment requires urgent rollback; and release authority sign-off from the accountable model owner confirming that all readiness conditions have been met.

Release notes documenting what changed between the current deployment and any prior version should be produced for every production release, not just initial deployments. When a model is updated through retraining on newer data, hyperparameter adjustment, or architectural change, the release notes should describe the nature of the change, the evaluation results from the updated model, and any differences in expected behavior compared to the prior version. This documentation supports post-incident investigation if the model update introduces a new failure mode and makes it easier for the governance committee to track the model's behavior over time.

Organizational readiness is distinct from technical readiness. A technically sound model deployed into an organization that lacks trained personnel to interpret its outputs, defined processes for acting on its recommendations, or awareness among affected staff that an AI system is operating in their workflow is an organizational risk regardless of its technical quality. Deployment readiness documentation should confirm that training for relevant staff has been completed, that communication about the model's capabilities and limitations has been provided to those who will rely on its outputs, and that the escalation path for questions or concerns about model behavior is known and accessible.

Diagram 8.7: Deployment Readiness Gate Checklist

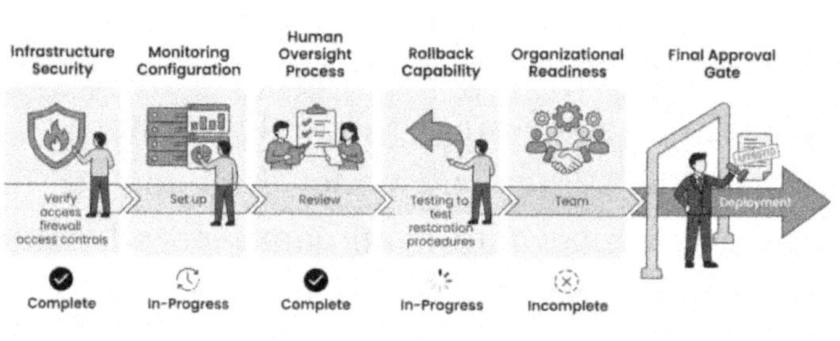

This checklist ensures all essential requirements are met before authorizing final system deployment.

8.8 Monitoring and Retirement

Production monitoring is the governance control that creates accountability for model behavior over time. Pre-deployment evaluation establishes that a model met its performance criteria at the time of deployment. Production monitoring determines whether those criteria continue to be met over time, as data distributions shift and the population served by the model changes. Models that are not monitored after deployment effectively operate without governance because the organization has no reliable mechanism to detect when performance degrades below acceptable levels.

The monitoring program for a production AI model should track multiple categories of metrics. Performance metrics measure whether the model continues to meet its acceptance criteria on current production data: accuracy, precision, recall, or the business metrics such as false denial rates or cost per correct prediction, as specified by the acceptance criteria. Data drift metrics measure whether the distribution of inputs arriving at the model has shifted significantly from the training distribution, because significant drift is a leading indicator of performance degradation even before accuracy measures show decline. Fairness metrics must be monitored continuously for models where fairness evaluation was part of the deployment requirements: a model that was fair at deployment may become unfair as the affected population changes. Operational metrics, such as latency, error rates, and availability, ensure that the model's technical performance meets the service levels required by the business processes that depend on it.

Alert thresholds for each monitoring metric should be defined before deployment as part of the deployment readiness documentation, not configured ad hoc after the model is live. Alert thresholds define the point at which a deviation in a metric triggers a human investigation or an automated response. Configuring thresholds too conservatively, too close to expected normal variation, leads to alert fatigue, causing responders to ignore alerts over time. Configuring them too permissively allows significant problems to develop before they are detected. Threshold calibration is an engineering and governance judgment that should be documented and reviewed periodically as the model's operating characteristics become better understood in production.

Retraining decisions must be governed with the same rigor as initial deployment. When monitoring indicates that a model's performance has degraded and retraining is proposed, the retrained model must pass through the same evaluation and validation requirements as the original deployment, not treated as a minor update exempt from governance controls. The common shortcut of treating retraining as routine maintenance and deploying retrained models without formal evaluation is a significant governance gap that has contributed to production AI incidents in which a retraining update introduced new biases or performance regressions.

8.8.1 Model Retirement and Decommissioning

Retirement is the end of the model's production life, and it deserves formal governance treatment rather than an informal shutdown. The retirement decision should be documented, recording the reason for retirement (replacement by a superior model, business process change, regulatory requirement, sustained performance failure), the date of retirement, and the disposition of all associated artifacts: training data, model files, inference logs, and documentation. The disposition of inference logs is particularly sensitive when they contain personal data, because applicable privacy regulations may govern their retention. Retirement documentation should confirm that the retention or deletion of the inference log has been handled in compliance with the data handling policy.

Residual data exposure in model parameters, the phenomenon where a retired model's parameters may encode information from its training data even after source data has been deleted, should be assessed at the retirement stage for models trained on personal data. The assessment does not require retraining the model, which is technically complex and may not be feasible for all architectures. It requires documenting the question and the organization's conclusion about residual exposure risk, so that a future inquiry can be answered with documented analysis rather than silence.

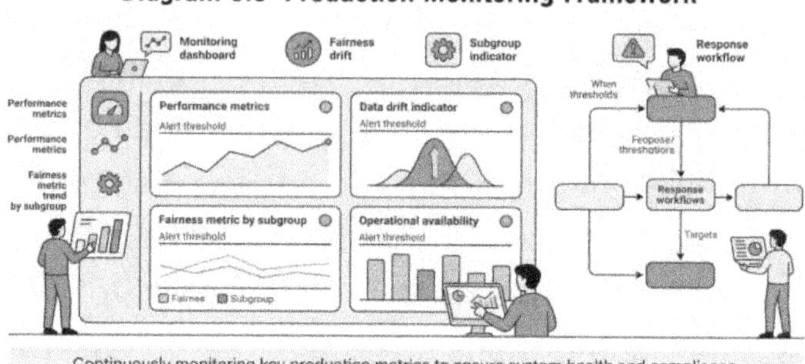

8.9 Manager's AI Lifecycle Checklist

The following behaviors and artifacts define a manager who has embedded governance into the AI development lifecycle. Review this list at the start of each new AI development project and at each stage-gate review.

Every AI development project begins with a documented use-case charter that includes a risk-tier classification approved by the accountable business owner. Development teams are operating in approved, managed environments with version control and experiment tracking in place. Training data for every production model includes complete provenance documentation, including source descriptions, legal basis for use, transformation records, quality assessment results, and version identifiers. Model cards are initiated during development and completed before deployment, not produced as afterthoughts during deployment preparation. Evaluation criteria are specified before evaluation begins, and any revision to acceptance criteria is approved by the business owner and documented. Fairness evaluation has been performed and documented for all medium- and high-risk models. The deployment readiness checklist has been completed, but not signed off on without verification covering infrastructure, monitoring configuration, human oversight processes, rollback capability, and organizational readiness. Production monitoring is in place with defined alert thresholds before the model goes live. Retraining events are treated as new deployments and must pass evaluation and validation requirements. Model retirement is documented, with confirmed disposition of training data, inference logs, model files, and an assessment of residual data exposure.

8.10 Embedding Governance in Engineering Culture

The most technically skilled AI development team will produce governance failures if they experience governance requirements as obstacles imposed from outside their work rather than as professional standards that are part of good engineering practice. Building a governance culture within data science and machine learning teams is therefore as important as designing the lifecycle governance process itself. Managers can influence this culture through specific, concrete actions: reviewing model cards and provenance documentation personally and providing feedback that demonstrates their importance; including governance compliance as a factor in project retrospectives and performance conversations; ensuring that governance requirements are included in sprint planning and project timelines rather than treated as overhead that must be completed on top of the "real" work; and recognizing teams that produce strong governance documentation, not just teams that achieve strong model performance.

The investment in lifecycle governance documentation pays dividends that extend well beyond regulatory compliance. Teams that maintain thorough experiment records build on their own prior work more effectively because they can reconstruct why earlier approaches were abandoned. Teams that produce complete model cards and validation documentation create institutional knowledge that survives personnel transitions. Teams that monitor their models in production learn faster from production experience than teams that deploy and forget. The well-designed governance process is not separate from good data science practice; it is a formalization of it.

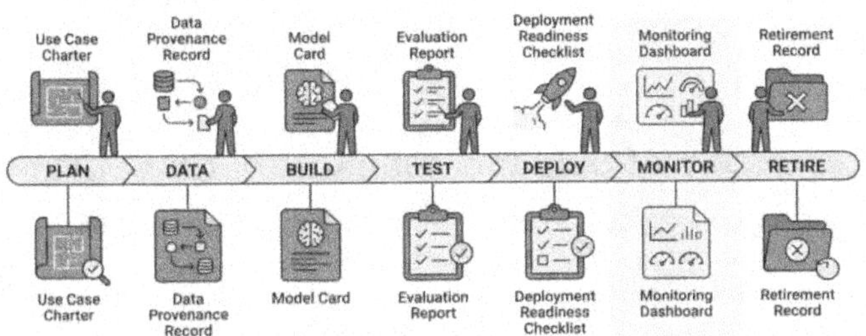

Diagram 8.9: Lifecycle Governance Documentation Map

This map visualizes the critical documentation artifacts tracked across each phase of the AI system governance lifecycle.

9 Purview and Privacy for AI Oversight

A large metropolitan transit authority deployed an AI system to optimize scheduling and service allocation across its network. The system used historical ridership data, anonymized trip records, fare transaction data, and third-party demographic data to predict demand patterns and recommend service adjustments. The project was sponsored by the operations department, procured as a vendor solution, and deployed without a formal privacy review. Eighteen months after deployment, an investigative journalist obtained the vendor's privacy terms under a public records request and discovered that the contract allowed the vendor to process fare transaction data, which, when combined with trip origin and destination records, could reconstruct identifiable travel patterns for vendor product improvement purposes. The authority had no record of having consented to this use or of having evaluated it. The subsequent regulatory inquiry found that the authority had not conducted a privacy impact assessment before deployment, had no lawful basis documentation for the secondary data processing, and had no mechanism for transit users to exercise their applicable data rights.

The incident did not involve a data breach. No unauthorized party obtained customer data. The privacy problem arose from decisions made during procurement and deployment about what data would be processed, by whom, and for what purposes, decisions that governance controls should have examined before the system went live. The transit authority's failure was not a technical failure but a governance failure: data governance and privacy oversight had not been applied to an AI system that processed personal data at a significant scale. This chapter addresses the governance controls that prevent exactly that failure.

9.1 Why Data Governance and Privacy Are Inseparable

AI systems process data in ways that are qualitatively different from conventional software. A conventional database application processes data records that were deliberately stored and can be precisely enumerated. An AI system processes data as input to mathematical operations that produce output predictions, recommendations, and classifications, which may implicitly encode information about individuals even when the system's designers did not intend to process any personal data. This characteristic of AI systems means that privacy considerations cannot be limited to the question of what personal data is stored; they must extend to questions about what personal data is processed during inference and what personal information is effectively retained in model parameters.

Data governance and privacy governance have historically been treated as separate disciplines within organizations. Data governance focuses on data quality, consistency, availability, and lifecycle management. Privacy governance focuses on compliance with applicable privacy laws and organizational privacy commitments. AI governance requires that these disciplines work in coordination, because the same data decisions that drive model performance, what data to collect, how to structure it, and how long to retain it, directly determine the organization's privacy risk exposure and compliance posture. A manager responsible for AI governance who treats data governance and privacy as someone else's concern will find that data and privacy failures consistently appear as AI governance failures.

The regulatory environment has amplified the stakes. Privacy regulations in most jurisdictions where AI-forward organizations operate, including the European Union's General Data Protection Regulation, the California Consumer Privacy Act and its amendments, and a growing body of sector-specific privacy requirements in healthcare, finance, and education, impose obligations on organizations that process personal data using automated systems. These obligations include requirements to have a lawful basis for processing, to limit processing to stated purposes, to protect data subjects' rights, to conduct assessments before certain types of processing, and to implement technical and organizational measures adequate to the risk. Ignorance of these obligations is not a defense, and most regulators now expect that organizations deploying AI systems have exercised judgment about their privacy implications before deployment.

Diagram 9.1: Data Governance and Privacy Oversight Integration

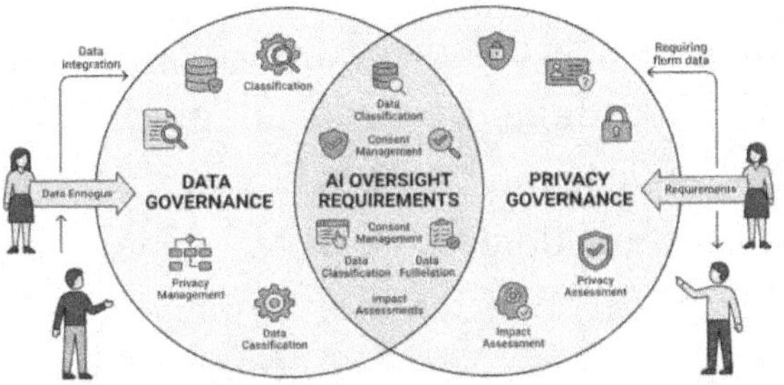

Integrated data and privacy governance establishes robust oversight framework for responsible AI deployment and compliance.

9.2 Data Governance Foundations

Data governance in the context of AI oversight begins with establishing clear accountability for the data that AI systems consume and produce. Every data set used in an AI system should have a named data steward responsible for its accuracy, appropriate use, and lifecycle management. The data steward is not a technical role exclusively; it is an accountability role that may be held by a business leader who understands the data's business context and can make decisions about its appropriate use. Where a data set is used across multiple AI systems, the data steward's accountability extends to coordinating governance requirements across all systems that share that data.

Data lineage, the documented path from data source through transformations to its use in a specific model, is a foundational data governance control for AI systems. Without lineage documentation, an organization cannot answer basic governance questions: If a source system changes its data collection practices, which AI models are affected? If a data processing error is discovered in a pipeline, which model versions were trained on the affected data? If a data subject exercises their right to erasure, which AI systems processed their data, and what is the impact of that erasure on each system? Lineage documentation enables these questions to be answered accurately and quickly rather than through expensive manual reconstruction.

Master data management principles apply to AI governance, even when the organization lacks a formal master data management program. The most directly relevant principle is the requirement that data used across multiple systems be governed with consistent definitions, quality standards, and classification. An AI system that receives customer age data from one system and demographic data from another may encounter inconsistencies between the sources, degrading model performance or introducing systematic errors in a specific subgroup. Data governance for AI requires not just managing each data source individually but also ensuring data consistency across the sources that feed into AI systems.

Data lifecycle management must account for AI-specific retention obligations. Conventional data retention schedules are typically driven by regulatory records requirements and business needs. AI introduces additional retention obligations: training data must be retained as long as the model it trained is in production, plus an investigation period afterward; inference logs may be required to support audit obligations; and model documentation, including data provenance records, may need to be retained as long as decisions made by the model remain subject to legal challenge. These AI-specific retention requirements should be integrated into the organization's overall data retention schedule rather than managed as an informal practice of the AI team.

Diagram 9.2: Data Governance Accountability Structure

Diagram 9.2: Data Governance accountability structure for onoin, employees ruon their data domain.

9.3 Data Classification

Data classification is the practice of assigning each data element or data set to a category that reflects its sensitivity, the harm that could result from its misuse or unauthorized disclosure, and the protection it therefore requires. For AI governance, data classification performs an additional function: it determines what data may be used in which AI development activities, what controls must be in place before classified data may enter a training pipeline, and what restrictions apply to inference logs that capture classified data as inputs.

A functional data classification scheme for AI governance typically uses three to five classification levels. A common four-level scheme uses the categories: public data, which may be freely shared and used without restriction; internal data, which is intended for internal use but whose exposure would not cause material harm; confidential data, which includes business-sensitive information whose exposure could cause competitive, operational, or reputational harm; and restricted data, which includes the most sensitive categories such as personal health information, financial account data, government identifiers, and other information whose unauthorized processing or disclosure could cause significant harm to individuals or the organization.

Each classification level should map to a defined set of AI governance controls. Restricted data used in AI training requires the most intensive controls: documented legal basis for use in AI, privacy impact assessment before training begins, access controls limited to individuals with a specific need, prohibition on use in development environments outside the production data governance perimeter, and audit logging of all access to training data. Internal data may require a lighter control set. Public data requires minimal control. Classification-based controls prevent the common failure mode in which development teams select training data entirely on its predictive value without considering the governance obligations triggered by the data's sensitivity.

Classification of AI system outputs is an underappreciated data governance requirement. A model trained on restricted data may produce output predictions, risk scores, and recommendations that are themselves sensitive. A credit risk score derived from protected financial information, a health risk prediction derived from clinical records, or an employee performance score derived from behavioral monitoring data are all AI outputs that inherit the sensitivity of the underlying training data. Output classification must consider not just the form of the output but its inferential content: an output that allows a sophisticated observer to reconstruct restricted source data is itself restricted, even if it does not directly present that data.

9.3.1 Special Categories and Protected Classes

Privacy regulations and anti-discrimination laws identify certain data categories that require heightened protection. These special categories include racial or ethnic origin, religious beliefs, health and medical information, sexual orientation and gender identity, biometric and genetic data, immigration status, and, in some jurisdictions, financial account data. AI systems that process special category data, including systems that use proxies or correlated variables that effectively function as special category data, trigger additional governance requirements in most regulatory frameworks.

For data classification purposes, the governance framework should specify how special category data is treated, typically by assigning it to the highest classification tier regardless of other factors. It should require that any AI system that processes or could infer special category data be identified at the scoping stage and governed accordingly. The risk is not limited to systems that obviously process health or biometric data. A workforce analytics model that uses commute patterns, schedule preferences, or leave records may effectively process health information if those patterns are correlated with medical conditions. Governance frameworks that define special category risks only in terms of directly collected protected data miss this inferential exposure.

Diagram 9.3: Data Classification Matrix for AI Use

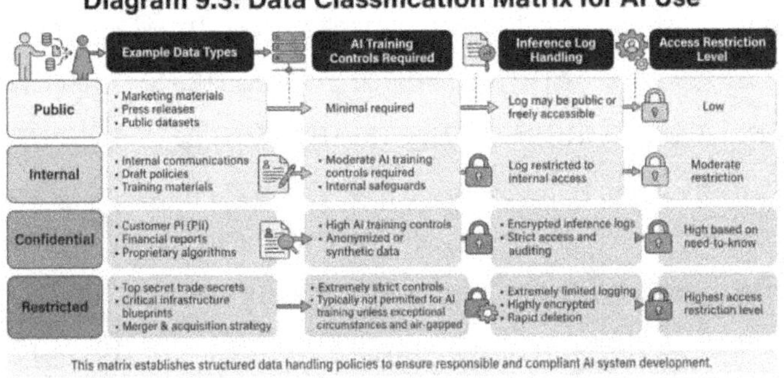

This matrix establishes structured data handling policies to ensure responsible and compliant AI system development.

9.4 Consent Models

Consent is one of the lawful bases under which personal data may be processed under privacy regulations. For AI systems, consent presents specific governance challenges because the processing of personal data in AI contexts, training, inference, and output storage, may be difficult to describe to data subjects in terms that allow genuinely informed consent. Explaining to a customer that their purchase history will be used as a feature in a recommendation model that shapes which products they see is relatively straightforward. Explaining that their browsing behavior will be used to fine-tune a large language model's personalization layer is less tractable.

AI governance programs need a structured approach to consent that distinguishes different types of AI data processing and determines which ones require consent and which may rely on alternative lawful bases. Under most privacy frameworks, consent is only one of several potential lawful bases. Legitimate interest, the organization's legitimate interest in using data to improve services, and contract performance using data to deliver a service the subject has contracted for, are commonly applicable lawful bases for many AI use cases. The critical governance requirement is that a lawful basis be identified and documented for every AI data processing activity before that processing begins, not identified retroactively when challenged.

Consent architecture for AI systems must address consent granularity: the degree of specificity at which consent is sought. Broad consent, "we may use your data to improve our services," is easier to obtain but is increasingly challenged by regulators as insufficiently specific to support AI training, which is a distinct and potentially unexpected use from the data subject's perspective. Granular consent, "we would like to use your data specifically to train our recommendation model, for the purpose of providing you with more relevant product suggestions," is more defensible but more operationally complex and more likely to be declined. The governance program must establish a consent architecture that is both legally defensible and operationally sustainable.

Consent withdrawal must be operationalized with the same rigor as consent collection. When a data subject withdraws consent for processing, the withdrawal must be honored prospectively; processing must cease; and, in some jurisdictions, prior processing that relied on the now-withdrawn consent must be addressed through data deletion or, where models were trained on the data, model retraining or retirement assessment. Governance documentation should specify the operational procedure for processing consent withdrawals in AI contexts, including the responsible parties, the time limits for action, and the documentation required to confirm compliance.

9.4.1 Alternatives to Consent

For AI systems where obtaining individual consent for data processing is impractical, large-scale analytics models, public-interest research systems, or systems that process data collected before AI use was contemplated, the governance program must establish and document an alternative lawful basis. Legitimate interest as a lawful basis requires a three-part test under most frameworks: documenting the legitimate interest pursued, demonstrating that the processing is necessary for that interest, and confirming that the interest is not overridden by the data subjects' interests and rights. This test must be performed and documented before processing begins, and the documentation must be updated if the processing purpose or scope changes materially.

Purpose limitation is the companion principle to lawful basis: data collected under one lawful basis for one stated purpose may not be repurposed for a different

AI use without either establishing a new lawful basis or determining that the new purpose is compatible with the original purpose. Compatibility is not a self-serving determination. "We collected this data for service delivery, and our recommendation model serves the user, so it's compatible," but a structured analysis that considers the link between purposes, the context of collection, the nature of the data, the consequences of the new processing, and applicable safeguards. Governance documentation should show this analysis rather than simply asserting compatibility.

Diagram 9.4: Consent and Lawful Basis Decision Framework

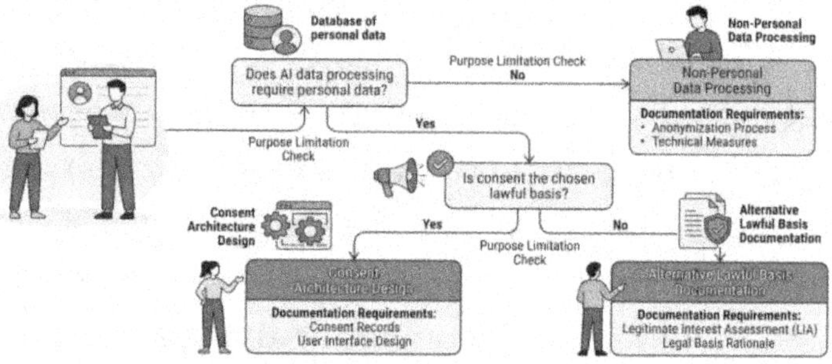

Systematic framework ensures GDPR compliance by validating data types and establishing clear legal justification for all AI activities.

9.5 De-identification Techniques

De-identification is the process of transforming personal data so that individuals cannot be identified from it, either directly or in combination with other available information. For AI governance, de-identification is a risk-mitigation control that allows organizations to use data for model development and training while reducing privacy risk and, in some cases, the scope of applicable privacy regulation. However, de-identification is not a binary state — data is not either identified or de-identified — and governance programs must be calibrated to the actual re-identification risk of the de-identification technique applied.

Pseudonymization — replacing direct identifiers such as names and account numbers with tokens or pseudonyms — is the most common de-identification technique. It substantially reduces the risk of casual identification but does not eliminate it, because the pseudonymized data may be linked to other datasets that restore the original identifiers. Pseudonymized data is, therefore, still personal data under most privacy frameworks and must be governed as such. However, the risk and associated obligations are generally lower than those associated with identified data. Governance documentation should specify the pseudonymization technique used, the key management practices that protect the link between pseudonyms and original identifiers, and the access controls that prevent unauthorized re-identification.

Aggregation and statistical disclosure control techniques reduce the risk of re-identification by ensuring that model inputs or outputs represent groups of individuals rather than specific persons. For AI training data, aggregation typically involves removing records for individuals who appear so rarely in the training set that their inclusion would make them identifiable. For AI outputs, aggregation involves ensuring that model outputs cannot be used to infer information about any specific individual in the training data. The governance requirement is not that a specific aggregation technique be used, but that the re-identification risk of the approach actually employed be assessed and documented.

Differential privacy is a mathematically rigorous technique that adds calibrated statistical noise to data or model parameters, providing a provable guarantee of the degree of privacy protection afforded. It is increasingly used in federated learning settings and in model training on sensitive data where organizations want to demonstrate to regulators that privacy has been protected by technical means, not just organizational policy. The governance documentation for models trained with differential privacy should specify the privacy budget used, the epsilon value that characterizes the privacy guarantee, and the tradeoff between privacy protection and model utility that was accepted.

9.5.1 Re-identification Risk Assessment

No de-identification technique provides absolute protection against re-identification. The practical protection depends on the technique's strength, the richness of the data that remains after de-identification, and the availability of external data sources that could be combined with the de-identified data to restore identifiers. Governance documentation for de-identified training data should include a re-identification risk assessment that considers all three factors and reaches a documented conclusion on whether the residual re-identification risk is acceptable given the use case's risk tier. This assessment should be performed by someone with technical expertise in privacy-enhancing technologies, not just asserted as a conclusion.

The legal status of de-identified data varies across jurisdictions and privacy frameworks. Under the GDPR, truly anonymous data that cannot be re-identified by any means reasonably likely to be used falls outside the scope of the regulation. Still, the standard for true anonymization is high, and regulators have been skeptical of claims that complex datasets have been truly anonymized. Under HIPAA, Safe Harbor de-identification and Expert Determination de-identification have defined standards. Governance programs should document which legal standard is being applied and why, rather than using the term "de-identified" without specifying the applicable standard.

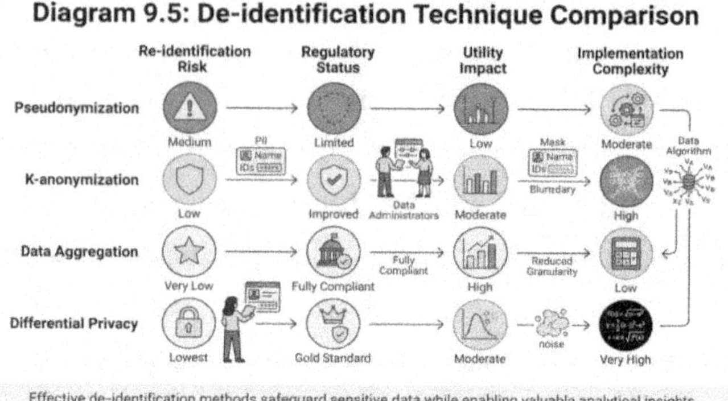

Diagram 9.5: De-identification Technique Comparison

Effective de-identification methods safeguard sensitive data while enabling valuable analytical insights.

9.6 Privacy Impact Assessments

A privacy impact assessment (PIA), also known as a data protection impact assessment (DPIA) under GDPR terminology, is a structured process for evaluating the privacy risks of a proposed data processing activity and identifying measures to reduce those risks before processing begins. For AI systems that process personal data at a significant scale or involve high-risk processing activities, such as systematic profiling, automated decision-making with legal or significant effects, or processing of special category data, PIAs are mandatory under several major privacy frameworks. Even where not legally mandated, PIAs are recognized as a best practice that demonstrates accountability and due diligence.

The PIA process for an AI system should cover: a description of the system and its data processing activities; a necessity and proportionality assessment confirming that the processing is necessary for the stated purpose and that a less privacy-invasive approach is not reasonably available; an assessment of risks to data subjects' rights and freedoms, including risks of discrimination, financial loss, reputational damage, unauthorized disclosure, and loss of data subject control; and measures proposed to address identified risks, with an assessment of whether those measures are adequate to bring residual risk to an acceptable level. For high-risk AI systems where the PIA identifies significant residual risk, most frameworks require consultation with the relevant data protection authority before deployment proceeds.

PIAs must be completed before deployment, not during or after. The purpose of a PIA is to inform decisions about whether and how to proceed with a processing activity, which means the assessment must occur while its findings can still influence decisions. A PIA completed after an AI system is already in production, as often happens when privacy review is a late-stage compliance step, can document risks and identify mitigations, but cannot prevent the risks that have already materialized and cannot function as a genuine deployment decision-making tool. Governance processes should make PIA completion a prerequisite for deployment readiness sign-off, not an optional step that follows deployment.

PIA documentation must be maintained and updated over time. The risks identified in a PIA at the time of initial deployment may change as the system evolves, new data sources are added, use cases expand, the affected population grows, or the regulatory environment changes. The governance program should specify the conditions that require a PIA review: significant changes to the system's data processing scope, deployment of the system in a new geographic or regulatory jurisdiction, discovery of a privacy incident affecting the system, or the passage of a defined time interval without review. PIAs that were current at deployment but have never been updated may not reflect the current processing activity of a mature AI system.

9.6.1 PIA for Automated Decision-Making

Automated decision-making using AI to make or substantially influence decisions that have legal or similarly significant effects on individuals triggers heightened PIA requirements under multiple privacy frameworks, as well as independent rights for data subjects to challenge such decisions, request human review, and contest the outcome. The PIA for an automated decision-making system must specifically assess the logic of the automated decision, the significance and likely consequences for the individual, and the measures in place to safeguard the subject's rights, including the human review process.

In practice, the definition of "substantial effect" for automated decision-making is broad. Decisions about credit access, insurance pricing, employment screening, content visibility on platforms, and access to government services all have significant effects under major regulatory frameworks. AI governance programs should identify all AI systems in the portfolio that make or substantially influence such decisions and ensure that each has a current PIA that specifically addresses the automated decision-making requirements applicable to it.

Diagram 9.6: Privacy Impact Assessment Process

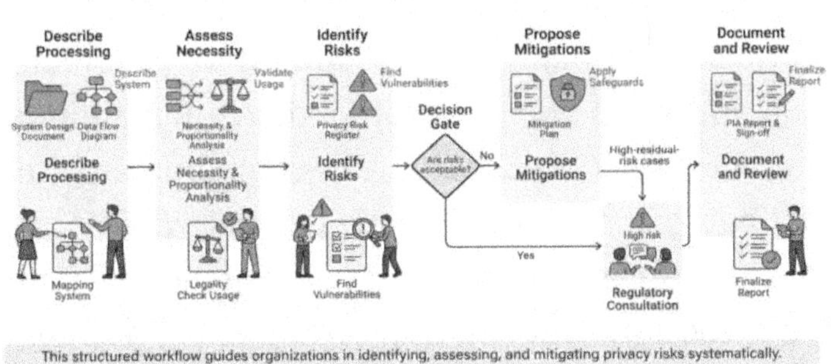

This structured workflow guides organizations in identifying, assessing, and mitigating privacy risks systematically.

9.7 Data Subject Rights

Data subject rights are legal entitlements that individuals have over the processing of their personal data. The specific rights vary by jurisdiction but typically include: the right to be informed about how data is processed; the right of access, to receive a copy of personal data and information about how it is being used; the right to rectification, to have inaccurate data corrected; the right to erasure, to have data deleted in certain circumstances; the right to restriction, to limit processing pending investigation of a dispute; the right to data portability, to receive data in a machine-readable format for transfer to another service; and the right to object to processing, including profiling. For AI systems, operationalizing these rights is more technically complex than for conventional data systems because AI systems process data in ways that are not always amenable to simple deletion or extraction.

The right of access applied to AI systems requires that the organization be able to identify, across all AI systems in its portfolio, what personal data about a given individual is being processed. This is technically straightforward for structured data in identified data sets, but becomes complex when personal data is distributed across multiple AI systems, when it has been transformed into features that are not directly mappable back to source records, or when it has been used to train model parameters that do not store personal data in any directly extractable form. Governance programs must document the scope of AI systems' personal data processing in sufficient detail to support fulfillment of access requests, including a maintained inventory of the personal data each AI system processes and in what form.

The right to erasure presents the most significant technical challenge for AI governance. Erasing a record from the training database does not erase the information that the record contributed to the model's learned parameters. The governance program must establish a policy for handling erasure requests when the personal data to be erased has been used in model training. Options include retraining the model without the data subject's records, which may be feasible for some architectures and impractical for others, applying machine unlearning techniques which are technically maturing but not yet universally applicable documenting that complete erasure from model parameters is not technically feasible and accepting the residual risk which may be acceptable under some regulatory frameworks as long as the source data is erased. Appropriate safeguards are in place or retiring the model and deploying a replacement trained without the data. Each option has different costs, and the governance program should specify a tiered approach based on the model's risk level and the availability of technical erasure mechanisms.

Profiling and automated decision-making rights the right not to be subject to a decision based solely on automated processing, or at a minimum to request human review of such a decision, require that AI systems be designed with human oversight processes that are meaningful rather than nominal. A human review process that rubber-stamps model outputs without genuine examination does not satisfy the right to human review. Governance documentation for systems that make automated decisions should specify the human reviewer's role, their authority to override the model, the criteria they use to review model outputs, and how they document their review decisions. This documentation demonstrates that human oversight is genuine rather than performative.

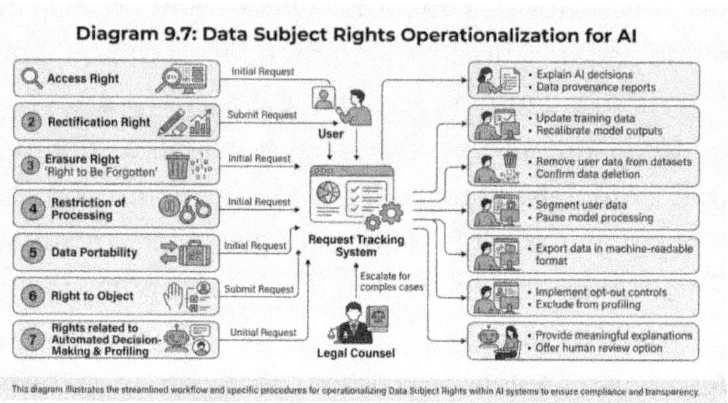

Diagram 9.7: Data Subject Rights Operationalization for AI

This diagram illustrates the streamlined workflow and specific procedures for operationalizing Data Subject Rights within AI systems to ensure compliance and transparency.

9.8 Balancing Utility and Privacy

The most persistent governance tension in AI data management is between model utility and privacy protection- the more data processed, the more varied and granular, the better the model performs, and the less data processed, the less individual-level data retained, the lower the risk. This tension cannot be resolved by prioritizing one value absolutely over the other. An AI governance program that always chooses privacy over utility will prevent the development of beneficial AI applications. A program that always prioritizes utility will accumulate privacy risks that eventually lead to regulatory exposure or loss of stakeholder trust. The governance challenge is to make principled, documented tradeoffs that optimize both values within defined risk tolerances.

Privacy by design is the principle that privacy protections should be engineered into AI systems from the start rather than added as compliance overlays after the system is built. Operationally, privacy by design means that data minimization — collecting and processing only the personal data that is genuinely necessary for the model's intended purpose — is a design requirement, not an optimization that can be addressed after the data architecture is established. It means that de-identification is applied as early in the pipeline as possible rather than only at the output. It means that retention limits are built into the data infrastructure rather than managed through periodic deletion campaigns. Privacy by design is a governance requirement that must be written into the development process, not left to individual practitioners' discretion.

Data minimization in AI development is technically non-trivial because the most common response to uncertainty about what features will be predictive is to collect everything and let the model determine relevance. This approach maximizes the data available for feature selection but also maximizes privacy risk. The governance requirement for data minimization does not preclude exploratory feature analysis. Still, it requires that features retained in the production model be justified by their predictive contribution, and that features that do not meaningfully contribute to model performance be removed from production data pipelines rather than retained indefinitely, even if they might be useful in future iterations.

Privacy-enhancing technologies, including federated learning, homomorphic encryption, and differential privacy, offer technical mechanisms to improve the privacy-utility trade-off. Federated learning allows models to be trained on data that never leaves the devices or systems where it is stored, providing privacy protection through data locality rather than de-identification. Homomorphic encryption allows computation on encrypted data, enabling analysis without decryption. These techniques are not yet universally applicable — they impose computational costs and may reduce model accuracy — but governance programs should track their development and include requirements for evaluating their applicability to high-risk AI use cases as part of the PIA process.

Transparency to data subjects about AI data processing is both a regulatory requirement and a trust-building investment. Privacy notices and AI system documentation that describe data processing in jargon-free, specific terms — rather than in broad boilerplate language — allow data subjects to make informed decisions about their data and reduce the risk of adverse reactions when AI use becomes public. Organizations that discover, through news coverage or regulatory inquiry, that they have been processing personal data in ways that subjects did not understand or would not have consented to, face reputational consequences that compound the direct regulatory risk. Transparency governance — specifying the content, accessibility, and accuracy of privacy notices for AI systems — is therefore not just a compliance function but a stakeholder relationship management function.

9.8.1 The Governance Tradeoff Record

When the governance program determines that a specific data use in an AI system involves a privacy risk that cannot be eliminated through de-identification or minimization but is justified by the system's benefits, that determination should be documented in what might be called a governance trade-off record. The record specifies the privacy risk, why the benefit of the data use justifies accepting that risk, what mitigations reduce the risk to the extent feasible, and who made the acceptance decision. This documentation serves multiple purposes: it creates an auditable record of deliberate risk acceptance rather than inadvertent exposure; it provides the basis for periodic review of whether the trade-off is still justified as circumstances change; and it demonstrates to regulators that the organization exercises genuine governance judgment rather than treating privacy as a checkbox requirement.

Diagram 9.8: Privacy-Utility Tradeoff Analysis Framework

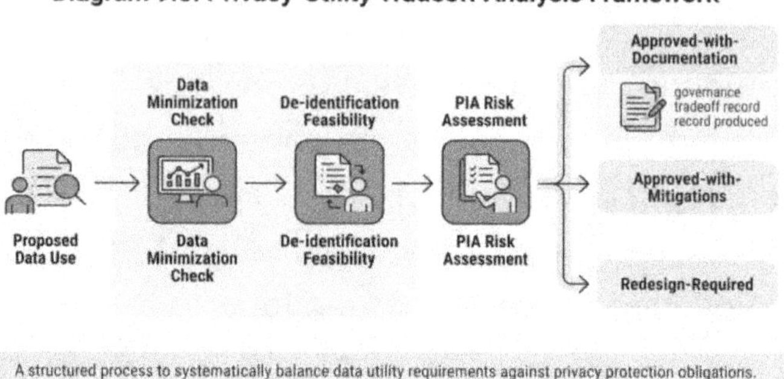

A structured process to systematically balance data utility requirements against privacy protection obligations.

9.9 Manager's Privacy and Data Governance Checklist

The following behaviors and artifacts define a manager who has operationalized data governance and privacy oversight for AI. Review this list before deploying any AI system that processes personal data.

Every AI system in the production portfolio has a named data steward responsible for the governance of its data inputs and outputs. Data lineage documentation exists for all production models and is sufficient to answer questions about which systems are affected by changes in any upstream data source. Data classification has been applied to all data used for AI training and inference, and classification-based controls restrict access to restricted-tier data. Consent or an alternative lawful basis has been identified and documented for every AI data processing activity before that processing began. De-identification techniques applied to training data have been assessed for re-identification risk, and the assessment is documented. A privacy impact assessment has been completed before deployment for every AI system that processes personal data, with the PIA reviewed and updated following material changes to the system. Data subject rights procedures are documented and operational for all AI systems, including procedures for access requests, erasure requests, and automated decision-making review requests. Privacy-by-design principles have been applied to new AI development projects, with data-minimization reviews conducted before the production pipeline is finalized. The governance program includes a process for producing governance tradeoff records when privacy risks are deliberately accepted and documentation of who made each acceptance decision.

9.10   Governance That Earns Trust Through Practice

Privacy and data governance for AI are not compliance obligations that can be satisfied by producing a set of documents and filing them with the appropriate regulatory authority. It is a practice that must be sustained through every data sourcing decision, every model development activity, every deployment, and every production data processing action. Organizations that approach it as documentation theater, producing PIA reports, consent notices, and data classification schemes that bear no relationship to what the organization actually does with data, will find that the theater provides no protection when regulators or litigants examine actual practice against stated policy.

Building genuine privacy governance capability requires investing in specific organizational functions: a privacy team with technical expertise in AI data processing, not just policy knowledge; data engineering practices that implement privacy-by-design requirements in actual pipelines, not just on paper; a data subject rights fulfillment operation that can respond to requests across AI systems in the required timelines; and a regular audit function that tests documented practices against actual behavior. These investments are substantial, but they are investments in the foundation of sustainable AI operations. Organizations that deploy AI at scale without this foundation are deferring privacy risk, not eliminating it, and the deferred risk tends to materialize at the most inconvenient possible moment.

The manager who governs data and privacy for AI well does not merely avoid regulatory penalties. They build AI systems that data subjects can trust, which is a competitive advantage in an environment where AI legitimacy is increasingly assessed based on how organizations treat the data of the people whose information powers their models. Trust earned through transparent, fair, and protective data governance is durable. Trust lost through privacy failures, even failures that were not technically illegal, is much more difficult to recover.

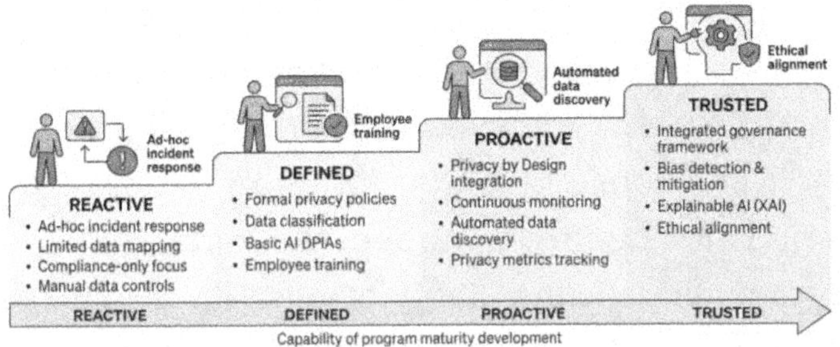

Diagram 9.9: Privacy Governance Maturity Model for AI

This model outlines the evolution from basic compliance to an ethical and integrated privacy governance framework for AI.

10 Overview of White House AI Bill of Rights

A mid-size financial services firm received a letter from a state regulator asking how its automated loan-scoring system complied with principles articulated in the White House Blueprint for an AI Bill of Rights. The compliance team forwarded the letter to IT. The IT director had read the Blueprint when it was released, but had filed it under "policy documents, non-binding." That filing decision now looked like a liability. Within three weeks, the firm had to produce evidence that its model was safe, that consumers could opt for a human alternative, and that automated decisions could be explained to affected individuals. None of those artifacts existed in a systematic form. This chapter prevents that situation.

The regulator's inquiry did not cite a statute that the firm had violated. It cited a series of principles that the firm was expected to have operationalized and invited the firm to demonstrate, through documentation, how it had done so. The firm could not demonstrate anything. It had deployed its model in a governance vacuum and discovered the vacuum only when external pressure forced a reckoning. The regulator was not unreasonable. The principles the inquiry referenced were publicly available, clearly stated, and had been cited in agency guidance documents for more than a year. The failure was organizational, not informational.

The White House Blueprint for an AI Bill of Rights, released in October 2022, is not a statute. It does not impose enforceable penalties on its own. Managers who stop at that observation miss the governance signal entirely. The Blueprint articulates five principles that regulators, litigants, and oversight bodies now routinely invoke when assessing whether an organization acted responsibly with AI. Federal agencies, including the CFPB, EEOC, and FTC, have each cited these principles in enforcement guidance. State legislatures in California, Colorado, and Illinois have codified subsets of them. The Blueprint is best understood as the consensus floor for what "responsible AI" means in the United States public sector and in regulated industries.

For managers, the practical implication is that building an AI governance program around only hard legal requirements is an increasingly fragile strategy. The distance between the Blueprint's principles and binding law continues to close. Organizations that treat the five principles as a program design framework rather than as aspirational prose will find that satisfying the more stringent legal requirements becomes largely automatic. The reverse is rarely true: compliance with a narrow sectoral rule does not automatically demonstrate the broader responsible-use posture that regulators and courts now expect.

This chapter translates each of the five Blueprint principles into concrete program requirements. It identifies the documentation, testing, and operational controls a manager must assemble to evidence compliance. It also situates the Blueprint within the broader ecosystem of national and international AI frameworks, such as the NIST AI Risk Management Framework, the EU AI Act, and the OECD AI Principles, so that managers can see how meeting one framework accelerates progress on others. The goal is not compliance with a document. The goal is governance maturity, expressed in artifacts and practices that exist independently of any particular regulatory instrument.

Diagram 10.1: Blueprint Principles Mapped to Governance Domains

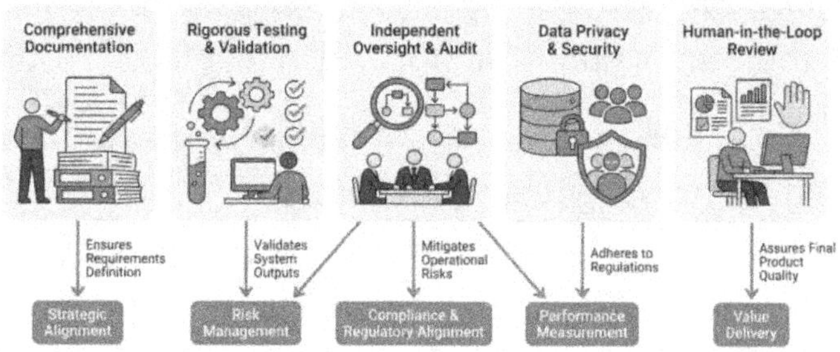

The Blueprint Principles provide a comprehensive framework that integrates seamlessly with established IT Governance Domains.

10.1   The Five Principles

The Blueprint organizes responsible AI around five principles: Safe and Effective Systems, Algorithmic Discrimination Protections, Data Privacy, Notice and Explanation, and Human Alternatives and Oversight. Each principle is accompanied by implementation guidance that describes what "applying" that principle looks like in practice. The guidance is written for a broad audience, including government agencies, private firms, and civil society. Still, the operational specifics are close enough to enterprise IT practice that a manager can build a control matrix directly from the text.

The five principles are not independent. They form an interdependent architecture. A system cannot meaningfully satisfy the Notice and Explanation principle if it has not first been evaluated for safety and discrimination risk because the notice must describe known limitations, and you cannot describe limitations you have not measured. Likewise, human oversight is only meaningful if the reviewer has sufficient context to make a genuine decision. That context comes from the transparency and documentation work required by other principles. Managers who treat the five principles as a checklist of independent tasks typically end up with documentation that does not cohere and gaps that surface during an audit.

The Blueprint is explicit that these principles apply to automated systems that affect individuals in consequential domains, such as credit, employment, education, healthcare, housing, criminal justice, and government services. Managers in other sectors should not conclude they are exempt. The principles reflect emerging social expectations about AI behavior that extend well beyond these enumerated domains. Any AI system that produces decisions or recommendations affecting individual customers, employees, patients, or citizens operates in the normative environment the Blueprint describes, regardless of whether a specific statute has codified its requirements.

10.1.1 Safe and Effective Systems

The first principle requires that AI systems be tested for safety and effectiveness before deployment and monitored continuously afterward. The Blueprint specifies that testing should be performed by independent evaluators where possible, that results should be documented, and that meaningful performance benchmarks should be established in advance rather than reverse-engineered after deployment. For a manager, this translates into three program artifacts: a pre-deployment test plan, a test results report, and a post-deployment monitoring protocol with defined alert thresholds.

The independence requirement deserves attention. For high-stakes systems that affect credit, employment, healthcare, or public benefits, the Blueprint calls for evaluation by parties without a direct stake in deployment success. In practice, this means that the team that built the model should not be the only one to validate it. A second-line risk function, an internal audit team, or an external vendor can fill this role. The key is documenting who performed the evaluation, what methodology they used, what findings they reached, and how those findings were resolved before deployment.

Safety and effectiveness extend beyond the model itself to the deployment system. A model that performs well in isolation may degrade significantly when integrated into a production pipeline with data quality variability, latency constraints, and edge cases that the test environment did not represent. Pre-deployment testing must include integration testing under realistic conditions, not just benchmark performance on a held-out dataset. The test plan should specify the assumptions about the production environment, the characteristics of the test data, and the stress conditions under which the system will be evaluated. Deviations between the test and production environments should be documented as limitations of the testing evidence.

10.1.2 Algorithmic Discrimination Protections

The second principle prohibits AI systems from producing discriminatory outcomes based on protected characteristics and requires proactive assessment of discrimination risk before deployment. The Blueprint specifies that this assessment must cover both disparate treatment (intentional differential handling based on protected class) and disparate impact (neutral criteria that disproportionately harm a protected group). This distinction matters because most AI discrimination issues arise from disparate impact: the model does not explicitly consider race or gender, but it uses proxies —

zip code, device type, application timing that correlate with protected characteristics.

Operationally, this principle requires a methodology for assessing fairness. The methodology must specify which demographic groups are evaluated, which fairness metrics are used (demographic parity, equalized odds, calibration), what thresholds trigger remediation, and how remediation decisions are documented. A manager who cannot produce a written fairness assessment methodology for each high-stakes AI system is exposed not only to Blueprint criticism but to Title VII, ECOA, and FHA liability, depending on the use case.

The choice of fairness metric is a substantive decision that should involve legal counsel and domain expertise, not just data science. Different fairness metrics are mathematically incompatible: a system that achieves demographic parity cannot simultaneously achieve equalized odds when base rates differ across groups. This is not a limitation to be engineered around; it is a fundamental constraint that requires a deliberate organizational choice about which fairness property is most important, given the use case. That choice must be documented, justified, and reviewed periodically as the deployment context evolves.

Diagram 10.2: Pre-Deployment Safety Testing Workflow

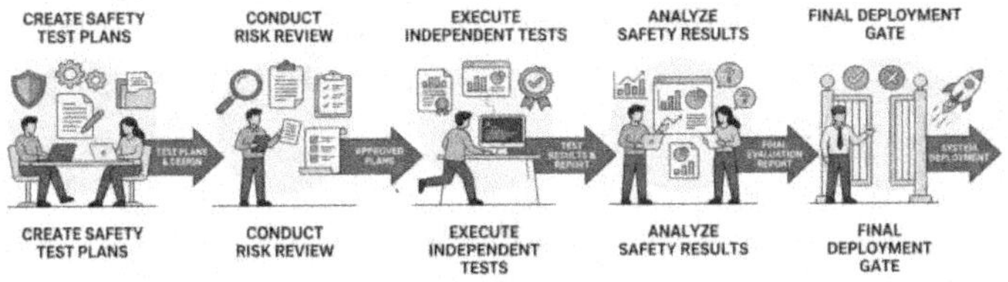

A structured process ensuring rigorous independent safety validation and review before production launch.

10.2   Transparency Requirements

Transparency is the connective tissue running through all five Blueprint principles. The Notice and Explanation principle makes transparency explicit: individuals subject to automated decisions must receive a plain-language explanation of how the system works, what data it uses, and what recourse is available. The explanation must be provided before or at the time of the decision, not after a complaint has been filed. This requirement has immediate implications for system design. If your AI system cannot generate an explanation that a non-technical person can understand, you have a program gap, not just a communication gap.

Managers should build transparency into the system from the model design phase, not bolt it on at deployment. This means selecting or requiring model architectures that support interpretability for high-stakes decisions. It means defining, at requirements time, what level of explanation is adequate for feature attribution, decision rules, confidence intervals, and testing that explanations are accurate and stable. An explanation that changes significantly with minor variations in input is not reliable; it is a post-hoc rationalization.

Transparency obligations extend beyond individual decision explanations to systemic transparency: organizations are expected to publicly disclose when they use AI systems to make or support decisions affecting individuals, to describe the categories of data used, and to provide accessible information about how individuals can seek human review. This systemic transparency is distinct from the individual explanation requirement, and both must be addressed. A comprehensive transparency program maintains a public-facing disclosure (often called an AI use registry or similar) alongside the individual explanation capability embedded in each deployed system.

10.2.1 Explanation Adequacy Standards

The Blueprint does not specify a single explanation methodology. Still, it does set a functional standard: the explanation must be "meaningful," which the guidance interprets as enabling the recipient to understand the basis for the decision and to identify potential errors. This is a higher standard than simply outputting feature importance scores. A meaningful explanation answers three questions for the affected individual: what was the primary basis for the decision, what factors would have changed the outcome, and how can the individual contest the decision if it is incorrect.

For managers, translating this into a control means requiring that every externally facing AI decision system produce a structured explanation object, not free-form text, but a structured record with defined fields, and that this object be stored alongside the decision record for the duration of the retention period. This structured record serves as the audit artifact that demonstrates compliance when a regulator or litigant demands evidence.

Explanation testing is a distinct quality assurance activity that most organizations have not built into their validation workflows. Testing explanations means verifying that they are accurate (the stated factors actually drove the decision according to the model internals), that they are stable (similar inputs produce similar explanations), that they are complete (material factors are not omitted), and that they are comprehensible (the explanation is understood as intended by representative samples of the affected population). Each of these properties requires a different testing methodology. Accuracy can be verified computationally. Stability can be tested through sensitivity analysis. Comprehensibility requires user research with non-technical participants.

**Diagram 10.3: Explanation Record Schema
for High-Stakes AI Decisions**

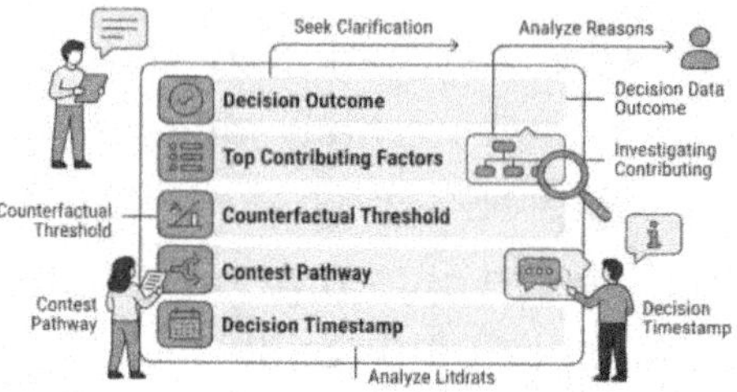

This schema defines essential data elements for documenting and interpreting high-stakes AI-driven outcomes transparently.

10.3   Safety and Effectiveness

Safety and effectiveness in the Blueprint's framing are not synonyms for accuracy. A highly accurate model can still be unsafe if it behaves erratically at distribution boundaries, is vulnerable to adversarial inputs, or has failure modes that cause disproportionate harm to specific populations. Effectiveness means that the system achieves its intended purpose in real operational conditions, not only on a benchmark dataset, and that the operational purpose is itself legitimate and clearly defined.

The Blueprint's implementation guidance for this principle calls for pre-deployment consultation with communities affected by the system, particularly in public-facing contexts. For enterprise IT managers, the closest analog is a structured business impact review that gathers input from business process owners, compliance, legal, and a sample of end users before deployment. The outputs of this review, documenting concerns and their resolution, form part of the safety evidence package.

Scope creep is a specific safety risk that the Blueprint addresses by requiring that systems be evaluated for their intended use and that deployment be limited to that scope. A model developed for internal workflow optimization, later repurposed for customer-facing decisions without re-evaluation, is not a safe deployment. The system has left the scope under which it was validated. This pattern is common in enterprise AI: a model that worked well in one context is extended to adjacent contexts where its assumptions no longer hold, often without a new validation cycle. The product charter and model card controls described in Chapter 13 are the mechanisms that prevent unreviewed scope expansion.

10.3.1 Ongoing Safety Monitoring

Post-deployment safety monitoring requires more than accuracy dashboards. The Blueprint specifies that systems should be continuously monitored for performance degradation, changes in discriminatory impact, and emerging safety issues. This requires a monitoring architecture that captures not only aggregate metrics but also distributional metrics, performance broken down by demographic group, input type, and time window. Aggregate accuracy can remain stable while performance for a specific subgroup degrades significantly; only disaggregated monitoring catches this pattern.

Monitoring also requires defined response protocols. When a metric crosses a threshold, who is notified? Within what time frame must a response plan be produced? What triggers a model suspension versus a model patch versus a human review overlay? These protocols must be written before deployment and tested in tabletop exercises. A monitoring system without response protocols is a warning system that no one can act on.

Safety monitoring should extend to user feedback channels. Affected individuals who experience apparent errors, biases, or unexpected behaviors in AI-assisted decisions should have a mechanism to report their observations, and those reports should flow into the monitoring function. Qualitative feedback from affected individuals often reveals safety issues before quantitative metrics detect them, particularly for edge cases that occur infrequently in aggregate but have a high impact for specific individuals. Building a feedback intake, routing, and analysis process is an underused yet high-value safety-monitoring control.

Diagram 10.4: Continuous Safety Monitoring Architecture

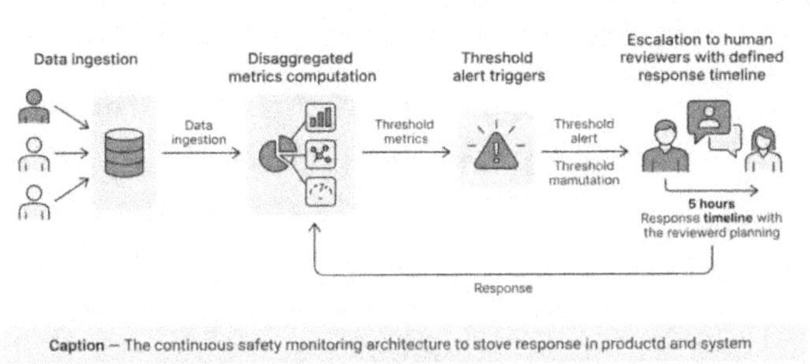

Caption — The continuous safety monitoring architecture to stove response in productd and system

10.4 Algorithmic Discrimination Protections

The discrimination protection principle has the most direct connection to existing legal exposure for most enterprise managers. Civil rights statutes in employment, lending, housing, and public accommodations have applied disparate impact theory for decades. The Blueprint extends this well-established legal concept explicitly to AI systems and calls for proactive assessment rather than reactive defense. The distinction between proactive and reactive matters enormously: a proactive fairness assessment conducted before harm occurs is a governance artifact; a fairness analysis conducted in response to a complaint is litigation preparation.

Managers should maintain a written fairness methodology document for each deployed AI system that affects individuals in any protected class context. This document specifies the protected classes evaluated, the fairness metrics and their mathematical definitions, the composition of the evaluation dataset, the acceptable disparity thresholds, and the remediation procedures triggered when thresholds are exceeded. This document must be version-controlled, reviewed annually, and updated when the model or its deployment context changes.

The Blueprint specifically calls for ongoing discrimination testing in production, not just at deployment. Fairness metrics measured on a pre-deployment test dataset reflect the characteristics of that dataset. They do not guarantee that fairness properties hold in production, where the input distribution may differ from the test distribution and the population of affected individuals may change over time. Annual re-evaluation with production data is the minimum cadence for high-stakes systems. Systems deployed in environments with known demographic change, growing markets, new geographies, and shifting customer populations may require more frequent re-evaluation.

10.4.1 Proxy Discrimination

Proxy discrimination — using features that correlate with protected characteristics without explicitly including those characteristics is the most common pattern of algorithmic discrimination and the most difficult to detect without deliberate testing. Common proxies include geographic features (neighborhood, zip code), behavioral features (shopping patterns, browsing history), and temporal features (application time, response latency). The presence of a proxy in a model does not automatically constitute discrimination; the question is whether its inclusion produces a disparate outcome that cannot be justified by legitimate business necessity.

The practical control for proxy discrimination is a feature audit: for each input feature, document whether it is correlated with any protected class at a statistically significant level in the training data. Features with high correlation require explicit justification for why this feature is necessary, and has the model been tested for disparate impact with and without it? This audit should be conducted during model development and repeated whenever the training dataset or feature set is updated.

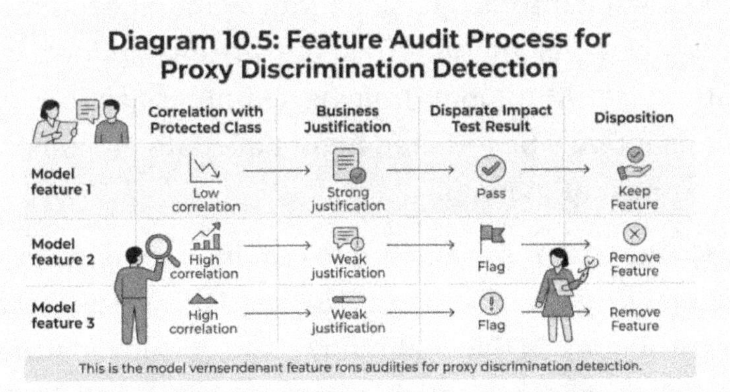

Diagram 10.5: Feature Audit Process for Proxy Discrimination Detection

10.5   Data Privacy

The Blueprint's data privacy principle echoes the privacy-by-design and data minimization requirements found in GDPR, CCPA, and other contemporary privacy regimes. Still, it applies specifically to the AI context. The central requirement is that data used to train, fine-tune, or operate AI systems should be limited to what is necessary for the stated purpose, that collection should be consented to or otherwise lawfully authorized, and that individuals should have a meaningful ability to contest the use of their data in automated decision-making.

For managers, the most operationally demanding aspect of this principle is the requirement for consent and authorization documentation. Many AI systems were built on data assembled before anyone thought carefully about the use of AI. That data may have been collected for a narrow purpose, transaction processing or customer service, and is now being used to train a model that makes decisions with a different scope. If the original consent or terms of service did not clearly authorize AI model training, re-authorization or re-consent may be required before the data can be used in a compliant manner.

The data privacy principle intersects with data subject rights in important ways. Individuals have the right, under GDPR and many state privacy statutes, to request deletion of their personal data. When that data was used in model training, deletion requests create a challenge: the trained model may encode information derived from the deleted individual's data even after the underlying records are removed. Techniques such as machine unlearning, which remove the influence of specific training examples from a trained model, are an active research area but are not yet widely deployed in production. Until robust unlearning tools are available, managers should document the organization's position on how deletion requests are handled in the context of trained models, and ensure that legal counsel has reviewed that position.

10.5.1 Data Minimization in Model Operations

Data minimization applies not only to training data but to the operational data pipeline. Every AI inference call that processes personal data should be reviewed to determine whether all the data provided is necessary for the inference. Many enterprise integrations pass full customer records to a model that only needs three or four fields. Passing unnecessary data increases privacy exposure, increases the risk of unauthorized disclosure, and creates retention obligations for data that served no function. A data minimization review examining what data is passed to each model at inference time and why is a low-cost, high-value control that most organizations have not implemented.

Inference data logging presents a related challenge. For audit and debugging purposes, it is often valuable to log model inputs and outputs. But logged inference data can be sensitive personal information subject to the same retention and access controls as source records.

Organizations that log inference data without a data classification, retention policy, and access control framework are accumulating a liability. Inference log governance, including who can access the logs, how long they are retained, the legal basis for retention, and when they must be deleted, should be part of the deployment design for any system that processes personal data.

10.6   Human Alternatives and Oversight

The human alternatives and oversight principle is the most operationally complex of the five because it requires not just documentation but functioning infrastructure. The Blueprint specifies that individuals subject to automated decisions must have access to a human alternative, a process where a person, not an automated system, reviews the decision. It further specifies that this alternative must be "meaningful," which the guidance defines as timely, consequential, and accessible. A human review process that takes six months, changes the outcome in 0% of cases, and requires a legal filing to access it is not a meaningful alternative.

For managers, building a meaningful human alternative requires answering several design questions before deployment: Who performs the human review? What information do they have access to? What authority do they have to override the automated decision? What is the response time commitment? How is the alternative communicated to affected individuals? How is the outcome of the human review documented? Each of these questions has a right answer and a wrong answer, and the wrong answers tend to cluster around two failure modes: the human reviewer lacks genuine authority, or the process exists on paper but lacks the resources to be operationally viable.

The Blueprint also addresses human oversight at the system level, not just the individual review level. Oversight means that qualified humans monitor the system's overall performance, can intervene to modify or suspend the system, and are empowered to act on monitoring findings. This is distinct from per-decision review. System-level oversight is the control that catches patterns. This model performs acceptably for most users but systematically fails for a specific subgroup that would not be visible in any individual review. Both levels of oversight are required for high-stakes systems.

10.6.1 Oversight Escalation Triggers

Not every AI decision requires a standing human review option; the Blueprint acknowledges proportionality. Low-stakes decisions with low error costs and easily reversible outcomes can operate with lighter-touch oversight. High-stakes decisions that affect employment, credit, benefits, healthcare, housing, or legal status require the full human alternative framework. Managers should maintain a documented decision classification that maps each AI use case to a stakeholder level and associates each stakeholder level with a specific oversight requirement. This classification is reviewed annually or when a new use case is added.

Escalation triggers for system-level oversight should be pre-defined and documented. Common triggers include: a monitoring metric crossing a threshold, an unusual volume of human review requests suggesting systemic issues, a change in the upstream data environment, published research findings about a model type or technique used by the system, and a regulatory development affecting the deployment context. When a trigger fires, the oversight process produces a documented assessment within a defined timeframe, with a disposition — continue unchanged, modify with re-validation, or suspend — and an owner for the required action.

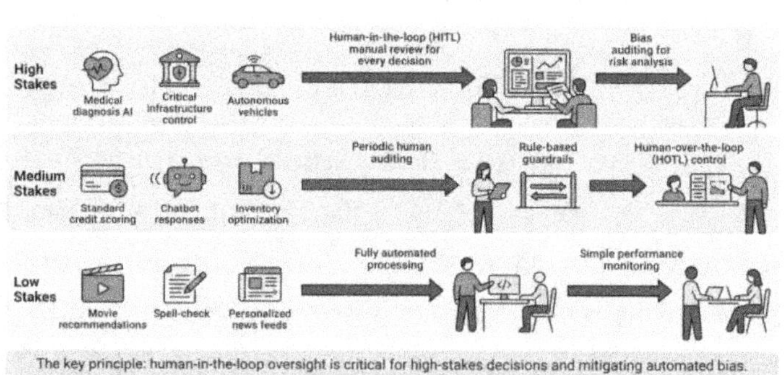

Diagram 10.6: Human Oversight Requirements by Decision Stakes Level

10.7   Building Evidence of Compliance

Evidence of compliance with the Blueprint's principles is not a single document. It is a portfolio of artifacts produced across the AI lifecycle, each of which speaks to one or more principles. Assembling this portfolio systematically rather than scrambling to reconstruct it under regulatory pressure is the difference between a mature AI governance program and an ad hoc response function. The following artifact types form the core compliance evidence portfolio for each deployed AI system.

The pre-deployment documentation package should include: a system purpose statement, a population impact description, a safety test plan and results report, a fairness assessment report, a data authorization review, an explanation adequacy assessment, and a human alternative design document. Each of these documents should name an owner, carry a version number, and reference the specific model version and deployment environment to which it applies. These documents constitute the baseline evidence package to be produced in response to a regulatory inquiry or legal challenge.

The ongoing compliance record adds to this baseline over time: monthly monitoring reports, annual fairness reassessments, records of human-review outcomes, incident reports, model-change records, and training completion records for staff with AI oversight responsibilities. Together, the baseline and ongoing records constitute the compliance evidence portfolio that demonstrates not only that the organization started correctly but that it maintained its governance posture over time.

Evidence portfolio management requires the same governance as the AI systems themselves. Documents must be stored in a manner that enables reliable retrieval, protects against unauthorized modification, is retained for the required period, and is associated with specific model versions. An evidence management process that relies on email threads, shared drives without version control, and undocumented storage locations will fail under audit scrutiny. The evidence repository is a governance tool, not an archive, and should be designed with the same care as any other compliance system.

10.7.1 Cross-Framework Mapping

The Blueprint's five principles map cleanly onto several other frameworks that enterprise managers are likely to encounter. The NIST AI Risk Management Framework organizes its controls around four functions: Govern, Map, Measure, and Manage, and each of these functions has direct correspondence to Blueprint principles. Safety and effectiveness map to NIST's Measure function. Transparency maps to Govern and Map. Discrimination protections map to Measure. Human oversight maps to Manage. Building compliance evidence for the Blueprint simultaneously produces the artifacts needed for an NIST-aligned program.

The EU AI Act, applicable to organizations deploying AI systems that affect EU residents, imposes binding requirements in several categories that align directly with Blueprint principles. High-risk AI systems under the EU AI Act require conformity assessments, technical documentation, human oversight measures, and transparency obligations, all of which correspond to Blueprint requirements. An organization that has built a Blueprint-compliant program for its U.S. operations will have completed most of the foundational work needed for EU AI Act compliance, significantly reducing the marginal effort required for multi-jurisdictional compliance.

The OECD AI Principles, adopted by 46 countries, provide a third reference point. Their five principles, inclusive growth, human-centered values, transparency, robustness, and accountability, map to the Blueprint's five with sufficient overlap that a single governance program can address both frameworks without duplicative effort. The practical implication for managers is that investing in Blueprint compliance is not a single-jurisdiction exercise. It is the foundation of a multi-framework AI governance capability that reduces compliance costs across an increasingly complex regulatory landscape.

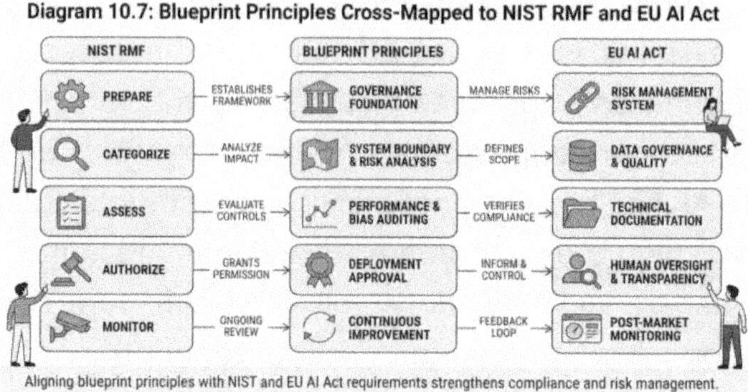

Diagram 10.7: Blueprint Principles Cross-Mapped to NIST RMF and EU AI Act

Aligning blueprint principles with NIST and EU AI Act requirements strengthens compliance and risk management.

10.8 Manager's Checklist

For each AI system that affects individuals in any high-stakes context, confirm the following are in place and documented before deployment and maintained post-deployment:

Pre-deployment: Written system purpose statement covering intended use, population affected, and decision stakes level. Safety test plan executed by a party independent of the development team, with results recorded and findings resolved. Fairness assessment covering all applicable protected classes with defined metrics, documented methodology, and acceptable thresholds, reviewed by legal counsel. Data authorization review confirming that training and operational data have a lawful basis for AI use, with particular attention to the scope of consent and minimization. Explanation adequacy assessment confirming that explanations meet the meaningful standard for the deployment context, tested with representative non-technical users. Human alternative design document covering reviewer qualifications, authority, response time, communication pathway, and resourcing.

Post-deployment: Monthly monitoring reports covering disaggregated performance metrics with threshold alerts and documented responses to any alert. Annual fairness re-assessment with comparison to baseline fairness measurement at deployment. Records of all human review requests and outcomes are reviewed quarterly for systemic patterns. Incident log for any safety or discrimination issues identified in operation, with root cause and resolution documentation. Version-controlled model update records with re-validation status, including scope comparison to prior version. Staff training records for all personnel with AI oversight responsibilities. Cross-framework compliance mapping is updated whenever a new regulatory requirement is identified.

10.9   Putting the Blueprint to Work

The White House Blueprint for an AI Bill of Rights is most useful when treated as a program design framework rather than a compliance checklist. Its five principles define the governance architecture that regulators, courts, and the public increasingly expect from responsible AI deployments. The manager who builds this architecture in advance — not in response to a regulator's letter — is the manager whose program can withstand scrutiny. The artifacts described in this chapter are not bureaucratic overhead; they are the evidence that demonstrates your organization takes its responsibilities seriously and has the operational discipline to back that claim.

Start with the highest-stakes systems in your portfolio. Assign an owner to each of the six pre-deployment artifact types. Set a deadline and a review process. The goal is not perfection on day one but a systematic, documented, continuously improving posture that leaves no AI system operating without evidence of its governance. When the next regulatory inquiry arrives, your response will be a set of documents, not a scramble.

The financial services firm from the opening scenario eventually completed its remediation. It produced the required documentation, retrained its staff, and implemented a human review process. It also spent three months doing it under regulatory pressure, at significantly higher cost than if the program had been built before deployment. The lesson is simple: Governance built reactively costs more and provides less protection than governance built proactively. The Blueprint gives you the framework. The discipline to act before you are required to is the governance leadership that the program depends on.

11 Generative AI and Large Language Model Controls

A large professional services firm deployed a generative AI assistant to help analysts draft client deliverables. Within two months the team identified three critical incidents: a model response that confidently cited a non-existent regulatory ruling, a prompt sequence that had induced the model to output internal salary benchmarking data that had been included in its context window by a poorly designed integration, and a content moderation gap that allowed a junior analyst to extract a competitor comparison that contradicted the firm's published positions. None of the incidents caused external harm, but each one exposed a governance gap that the firm had not anticipated when it approved deployment. The model had passed its accuracy benchmarks. It had failed its operational governance requirements.

The post-incident review found that the firm had applied its standard ML governance checklist to an LLM deployment. The checklist asked whether the model had been validated, whether performance metrics had been established, and whether a monitoring plan existed. All three answers were yes. But the checklist had no questions about prompt injection, context window data hygiene, or adversarial testing. The governance framework had been designed for a different class of system, and the gap between the framework and reality led to three incidents before anyone recognized it.

Generative AI and large language models introduce a category of governance challenge that does not fit neatly into the control frameworks designed for traditional ML systems. Classical discriminative models, classifiers, regressors, and rankers have bounded output spaces. You can enumerate the categories they produce and test each one. LLMs produce unbounded natural language output. Their behavior is highly sensitive to input context, instruction framing, and the composition of their context windows. They can be induced to behave in ways their developers did not intend through carefully crafted inputs, a class of attack known as prompt injection. They can generate plausible-sounding falsehoods with high confidence, in ways that are difficult for a non-expert reviewer to detect in real time.

These properties do not make LLMs ungovernable. They do require a different governance architecture. The controls needed for LLM deployment are more operational and more continuous than those for classical models. They include guardrail systems, output monitoring, human-review workflows, and vendor-accountability frameworks that lack direct analogs in the classical ML governance playbook. Managers who attempt to govern LLMs using only the tools designed for classical models will systematically miss the failure modes that matter most.

This chapter provides the governance architecture for generative AI and LLM deployments. It covers the specific failure modes that require dedicated controls, the operational mechanisms available to managers, and the vendor assessment criteria that determine whether a third-party LLM provider is governable at all. The chapter is organized around the governance lifecycle: understand the risks, implement controls, monitor in production, and assess vendors systematically.

Diagram 11.1: LLM Governance Failure Mode Taxonomy

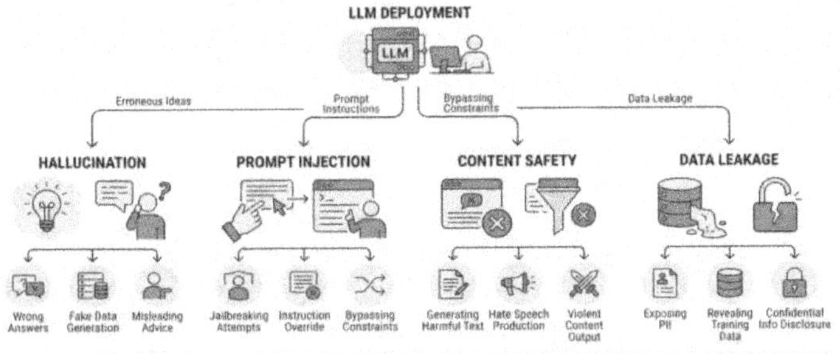

A visual classification framework for identifying and mitigating risks in Large Language Model governance failure modes.

11.1 LLM Governance Challenges

Large language model governance begins with an accurate threat model. The four primary failure modes, hallucination, prompt injection, content safety failures, and unintended data disclosure, each require different control mechanisms and have different likelihood and impact profiles depending on deployment context. A manager who conflates these failure modes will deploy controls that address the wrong risks.

Hallucination refers to model outputs that are confidently stated but factually incorrect or entirely fabricated. LLMs generate statistically plausible text; plausibility is not the same as accuracy. In low-stakes creative applications, hallucination is an inconvenience. In professional, medical, legal, or financial contexts, a hallucinated regulatory citation or a fabricated clinical finding can cause material harm. The governance response to hallucination is not to prohibit LLM use in these contexts, but to implement retrieval augmentation, citation requirements, and human verification workflows calibrated to the stakes of the use case.

Prompt injection is the LLM equivalent of SQL injection. An adversarial actor, or more commonly, a poorly designed integration, introduces instructions into the model's context that override its original system instructions or induce it to perform unintended operations. The risk is especially acute in agentic LLM deployments, where the model is authorized to take actions, execute code, query databases, and send messages based on its interpretation of inputs. A model that can be induced to execute arbitrary code through a manipulated input is a high-severity vulnerability, not a model limitation.

Unintended data disclosure occurs when the model outputs information from its context window, its training data, or connected systems that it was not intended to reveal. This includes personal data about individuals, proprietary business information, confidential employee data, and sensitive operational details. The risk arises because LLMs are designed to be helpful and responsive; they produce outputs that satisfy user requests, and if a request, deliberate or incidental, is satisfied by disclosing sensitive information, the model will often do so unless explicitly instructed otherwise.

11.1.1 Context Window Risk

The context window, the full set of text the model processes at inference time, is both the source of LLM capabilities and a major surface for governance risk. In enterprise deployments, context windows routinely contain system instructions, retrieved documents, user conversation history, and API-injected data. Each of these components can contain sensitive information. If the model is instructed, induced, or inclined to reproduce context window content in its outputs, that content becomes visible to the user and potentially logged, transmitted, or stored in ways it was never authorized for. Context window hygiene, controlling what is inserted into context, at what sensitivity level, and under what authorization, is an underappreciated governance control.

Context window governance requires a data classification layer applied to every integration that injects content into the model's context. Documents classified above a defined sensitivity threshold should not be inserted into context windows that are accessible to users without the corresponding clearance. System instructions containing operational secrets, API keys, internal pricing, and competitive strategy should be stored in a protected infrastructure and injected through secured channels rather than hardcoded in plaintext prompts. Context window content should be logged with the same data handling requirements as the source data.

11.2   Prompt Governance

Prompt governance is the systematic management of the instructions, templates, and input patterns that shape model behavior. In a minimally governed deployment, prompts are written by individual developers or users and evolve organically. In a governed deployment, system prompts are version-controlled artifacts, reviewed before deployment, tested against defined behavioral requirements, and changed through a formal change management process. The difference in outcomes between these two approaches is large.

The system prompt, which constitutes the preliminary set of instructions provided to the model before user engagement, serves as the chief regulatory tool for the deploying organization. It establishes the model's role, sets behavioral boundaries, defines output format requirements, specifies topics the model should not address, and instructs the model on handling edge cases. A well-crafted system prompt can substantially reduce the frequency of harmful, off-topic, or non-compliant outputs. It does not eliminate them. Prompt injection and distributional edge cases will always produce outputs that violate the system prompt's intent. This is why prompt governance is necessary but not sufficient.

User prompts and inputs provided during interactions pose a different governance challenge. Unlike system prompts, which the deploying organization controls, user prompts are externally generated and can contain adversarial content, sensitive data, or instructions that attempt to override system behavior. Input screening, reviewing user prompts for policy violations, sensitive data, and injection attempts before they reach the model, is the first-line control for user prompt risk. This screening can be automated (classifier-based) or policy-based (pattern matching for known attack signatures), and is most effective when it operates on the structured prompt components rather than the free-form text alone.

11.2.1 Prompt Version Control

System prompts must be managed with the same rigor as production software. Each change to a system prompt should be documented in a change record, tested against a defined behavioral test suite, approved by the system owner, and deployed with a rollback plan in place. Prompt changes are not configuration changes; they alter model behavior in ways that can be as significant as a code change, and they deserve equivalent scrutiny. Maintaining a prompt version history also creates the audit evidence needed to demonstrate that the model's instructed behavior at any point in time was reviewed and approved.

Behavioral test suites for system prompts should include both positive tests (the model produces the expected response to normal inputs) and negative tests (the model declines to produce harmful responses to adversarial inputs). Negative tests should cover the specific failure modes relevant to the deployment context: for a customer-facing legal information system, negative tests include attempts to elicit specific legal advice; for an HR assistant, they include attempts to extract confidential personnel information. The test suite is a living document, expanded as new attack patterns are discovered in production.

Diagram 11.2: Prompt Governance Change Management Process

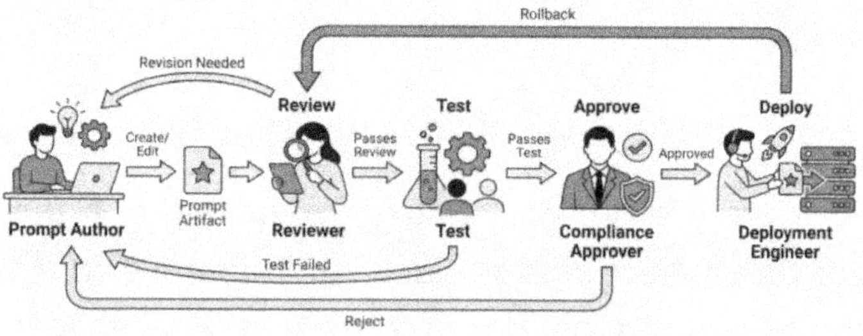

A structured workflow ensures prompt changes are validated, compliant, and deployed safely, with rollback capability.

11.3   Hallucination Mitigation

No current LLM architecture eliminates hallucination. The governance task is to reduce its frequency through system design and to detect and contain its impact through operational controls. The most effective architectural mitigation is retrieval-augmented generation (RAG). Rather than relying on the model's parametric memory, the system retrieves relevant documents at inference time and instructs the model to base its response on the retrieved content, with citations. This approach reduces hallucination frequency substantially in knowledge-intensive tasks because the model has access to authoritative source material and can be required to ground its outputs in it.

RAG is not a complete solution. Models can still fabricate details not present in retrieved documents, misattribute citations, or selectively ignore retrieved content that contradicts their parametric tendencies. The governance complement to RAG is citation verification: requiring the system to include source references with each factual claim and building a verification layer, automated or human, that confirms the cited source actually supports the claim. For high-stakes outputs, automated citation verification using a separate retrieval query is technically feasible and increasingly standard practice.

Confidence calibration is a complementary mitigation. A well-calibrated model expresses uncertainty accurately when it says it is 90% confident; it is correct 90% of the time. Models that are systematically overconfident, stating uncertain claims with high apparent certainty, are more dangerous in practice than models that express appropriate uncertainty, because users and downstream systems cannot apply appropriate skepticism. Calibration evaluation is a standard technique from classical ML that is not yet routinely applied to LLM deployments. For high-stakes applications, calibration testing comparing stated confidence to empirical accuracy across a representative evaluation set should be part of the pre-deployment validation.

11.3.1 Human Verification Thresholds

For deployments where hallucination risk is significant, professional advice, regulated information, medical content, and legal analysis organizations should define verification thresholds: categories of output that require human review before they are acted upon or transmitted externally. These thresholds should be defined by use case, output type, and stakeholder level, documented in the system's operational runbook, and enforced through workflow controls that route qualifying outputs to a review queue. The threshold definitions should be revisited quarterly based on the hallucination patterns observed in production logs.

Human reviewers in verification workflows must be qualified to assess the accuracy of the outputs they review. A reviewer who cannot assess whether a legal citation is accurate cannot provide meaningful verification for a legal information system. The verification workflow design must include reviewer qualification requirements, access to reference material, and escalation paths for outputs when the reviewer is uncertain. Verification workflows that produce reliable assessments require ongoing quality assurance of the reviewers themselves, tracking reviewer disagreement rates, calibration against known-correct outputs, and periodic competence assessments.

Diagram 11.3: RAG Architecture with Citation Verification Layer

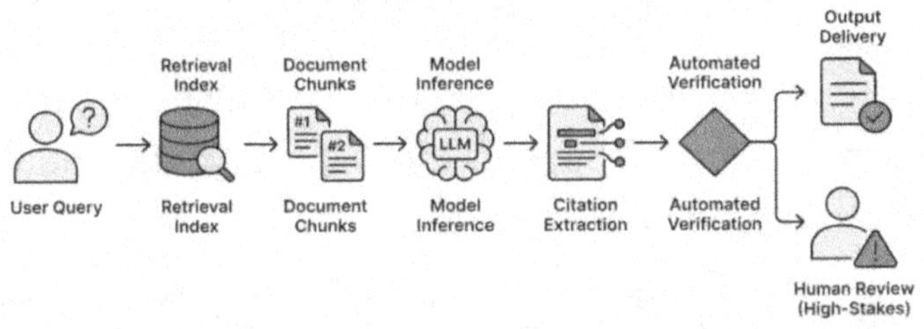

A robust RAG system ensuring accurate answers with automated citation verification and conditional human oversight.

11.4   Content Safety Controls

Content safety controls address the risk that an LLM produces outputs that are harmful, illegal, offensive, or otherwise contrary to organizational policy. These controls operate at multiple layers: the model's base training (handled by the model provider), the system prompt instructions, moderation classifiers applied to inputs and outputs, and human escalation workflows for edge cases. A well-designed content safety architecture uses defense in depth — no single layer is expected to catch everything; subsequent layers catch failures at earlier layers.

Input moderation classifiers screen user inputs for categories of harmful content, harassment, explicit material, and instructions for illegal activities before they reach the model. Output moderation classifiers screen the model's responses before they are delivered to the user. Major model providers offer these classifiers as API features; some organizations deploy additional classifiers tuned to their specific content policy requirements. The key governance requirement is that these classifiers are tested for both false positive rate (legitimate content blocked) and false negative rate (harmful content passed through), and that both rates are monitored in production.

Content policy for enterprise deployments must be written before deployment, not derived retrospectively from incident patterns. The policy specifies the categories of content the system will not produce, the categories of user input it will not process, the handling for edge cases and ambiguous requests, and the escalation pathway when a user disputes a content decision. This policy should be reviewed by legal counsel and compliance before deployment, updated when organizational policy or regulatory requirements change, and communicated clearly to users through the system's terms of use or user documentation.

11.4.1 Policy-Specific Content Rules

Content safety for enterprise deployments requires policy-specific rules beyond the general categories of harmful content. Financial services firms need rules around investment advice, suitability determinations, and regulatory disclosures. Healthcare organizations need rules around clinical recommendations, diagnostic statements, and treatment instructions. Legal services firms need rules around attorney-client privilege and unauthorized practice of law. Each of these rule sets must be specified in the system prompt, enforced by output classifiers tuned to the domain, and tested with domain-specific adversarial inputs before deployment. Generic content safety controls designed for consumer applications will not catch domain-specific policy violations.

Domain-specific content rules require domain expertise to design and to test. A content policy for a healthcare LLM that does not involve clinical informaticists and patient safety professionals is likely to have gaps. A financial services content policy that securities compliance counsel has not reviewed may prohibit the wrong things or fail to prohibit the right ones. The investment in domain expertise at the content policy design stage is far less costly than the reputational and regulatory exposure from deploying a system that violates domain-specific standards.

Diagram 11.4: Layered Content Safety Architecture

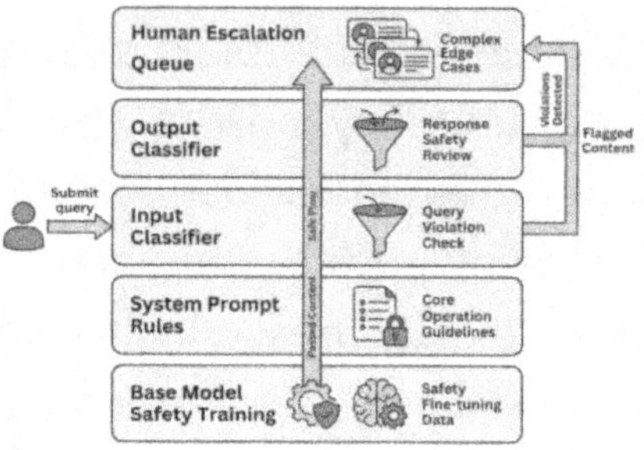

Greening content. make concise. pech that are unicoatifty to stono prionar, safety and conoil safety.

11.5 Watermarking and Provenance

Watermarking and provenance controls address a governance challenge that has no analog in classical ML: distinguishing AI-generated content from human-generated content and tracing it back to the specific model, version, and session that produced it. As AI-generated text becomes indistinguishable from human text at the surface level, organizations face liability for AI outputs mistakenly attributed to human authors, and regulators increasingly require disclosure of AI authorship in regulated communications.

Cryptographic watermarking, embedding a statistical signature in the model's output probability distributions, is the most technically robust approach to AI content attribution. Several research implementations exist; commercial deployments are emerging. For most enterprise managers today, the practical alternative is metadata-based provenance: logging each LLM inference call with the model version, session identifier, input hash, and output hash, creating an auditable record that a specific output was generated by a specific model under specific conditions. This log-based provenance does not embed attribution in the content itself, but it satisfies the governance requirement of traceability for audit purposes.

Provenance metadata should be stored in a tamper-evident system and retained for the same period as the decisions or communications to which the AI-generated content contributed. For financial communications, this typically means retention periods matching those for the underlying transaction records. For healthcare content, it means alignment with medical record retention requirements. For employment-related content, it means retention through any applicable limitations period for discrimination claims. The provenance log is not merely a debugging tool; it is an audit artifact.

11.5.1 Disclosure Requirements

Several jurisdictions now require disclosure of AI authorship in regulated contexts. The EU AI Act requires disclosure when AI interacts with natural persons in real time unless the context makes it obvious. FTC guidance on endorsements has been extended to AI-generated testimonials. Several state laws require disclosure of AI-generated political communications. Managers should maintain a disclosure matrix: for each LLM deployment, what disclosure obligations apply, where and how disclosure is delivered, and who is responsible for ensuring disclosure happens. This matrix is a living document updated as regulatory requirements evolve.

11.6 Evaluation Metrics

Evaluating LLM systems requires a different metric portfolio than classical ML. Accuracy in the traditional sense, comparing model output to a labeled ground truth, is applicable in constrained tasks like classification or extraction, but does not capture the quality dimensions that matter most for generative outputs: factual accuracy, coherence, instruction-following, and absence of harmful content. Managers must define evaluation frameworks before deployment that specify what metrics will be used, how they will be measured, and what thresholds must be met before a system is approved for production.

The core evaluation dimensions for enterprise LLM systems are: faithfulness (does the output accurately reflect the sources or context provided), answer relevance (does the output address the question asked), context recall (does the output use all relevant information available), safety (does the output comply with content policy), and latency (does the system respond within acceptable time limits under production load). Each of these dimensions requires specific measurement approaches. Faithfulness and answer relevance can be measured with automated LLM-asjudge frameworks. Safety requires both automated classifiers and human red-teaming. Latency requires load testing under representative conditions.

Benchmark selection for LLM evaluation requires care. Published benchmarks measure specific capabilities in specific conditions. An LLM that performs well on a standard reasoning benchmark may perform poorly on your specific enterprise use case, particularly if your use case involves domain-specific knowledge, proprietary data formats, or interaction patterns not represented in public benchmarks. Enterprise evaluation must include use-case-specific test sets and real or realistic examples of the tasks the system will perform, as the primary evidence. Published benchmark performance is supplementary context, not the primary basis for deployment approval.

Diagram 11.5: LLM Evaluation Framework with Metric Dimensions

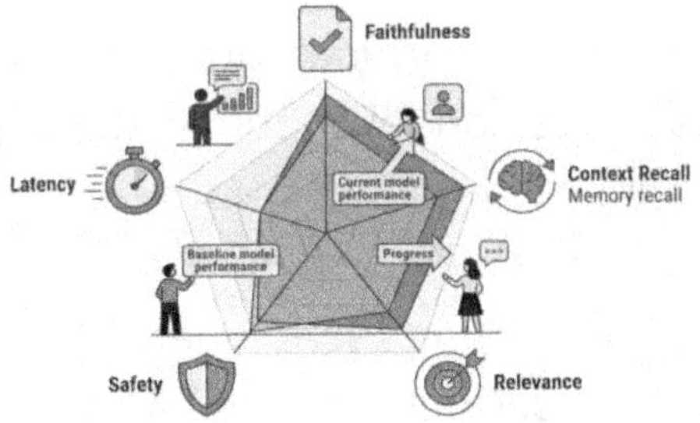

Metric dimensions visualize performance comparison for LLM optimization.

11.6.1 Red-Teaming Requirements

Red-teaming, systematic adversarial testing by a team attempting to elicit harmful, non-compliant, or unexpected behaviors, is a required governance control for any LLM deployed in a context where harmful outputs would have material consequences. Red-teaming should be conducted before deployment, repeated after significant model or prompt updates, and documented with specificity: which categories of behavior were tested, which prompting strategies were used, which harmful outputs were elicited, and which mitigations were implemented in response. A red-team report that says "no issues found" without specifying methodology and scope is not useful evidence; it is documentation theater.

Red-team scope for enterprise deployments should be tailored to the specific deployment context rather than replicating generic LLM safety evaluations. A financial services LLM red-team tests for regulatory violations, fiduciary breaches, and data disclosure. A healthcare LLM red-team tests for clinical recommendation overreach and patient safety failures. The red-team personnel should include domain experts who understand what a harmful output looks like in context, not just security personnel who understand adversarial prompting techniques. The combination of domain expertise and adversarial technique produces more relevant findings than either alone.

11.7 Vendor Assessment

Most enterprise LLM deployments rely on third-party model providers. The governance implications of this dependency are significant and underappreciated. When an organization deploys an external LLM, it is operationally dependent on the provider's training decisions, safety mitigations, infrastructure reliability, and data handling practices. If the provider changes its model without notice, it is a common practice, since many providers release new model versions frequently, and the organization's tested behavior guarantees may no longer hold. If the provider stores or uses inference data for training, the organization's data privacy commitments to its customers may be breached. Vendor assessment for LLMs must specifically address these risks.

The LLM vendor assessment framework covers five domains: model transparency, data handling, change management, security, and contractual protections. Model transparency asks: Does the provider publish documentation of training data sources, safety evaluation methodology, and known limitations? Data handling asks: where is inference data stored, is it used for training, what retention period applies, and what data processing addendum governs the relationship? Change management asks: how are model version updates communicated, what notice period is provided, and is it possible to pin to a specific version for production stability? Security asks: What certifications does the provider hold? What is the incident notification SLA? How is access to inference logs controlled?

Provider concentration risk deserves specific attention. Organizations that have deployed multiple LLM-powered applications on a single provider's infrastructure are operationally dependent on that provider's reliability, pricing, and policy decisions. A pricing change, a terms-of-service update, or an outage has portfolio-wide impact. Governance programs should include a provider dependency assessment that identifies single-provider concentrations and evaluates the feasibility of mitigation strategies, alternative providers, model portability, and fallback to classical systems for each high-stakes deployment.

11.7.1 Contractual Minimum Requirements

No LLM vendor engagement should proceed without a data processing addendum (DPA) that explicitly prohibits the use of inference data for model training without separate written consent, specifies data retention and deletion timelines, names the jurisdictions where data is processed, and provides an audit right. Additionally, the master services agreement should include: model version change notification requirements with a minimum notice period, a right to test new model versions before they are forced into production, availability SLAs with credit provisions, and a security incident notification requirement with a defined response time. Vendors who refuse these provisions should be evaluated with heightened scrutiny or declined for high-stakes deployments.

Diagram 11.6: LLM Vendor Assessment Scorecard

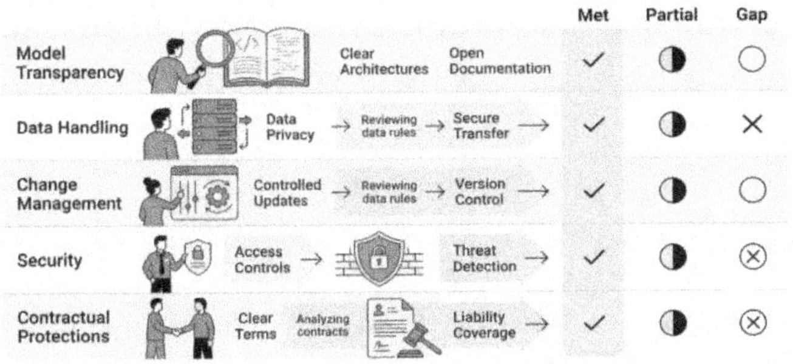

		Met	Partial	Gap
Model Transparency	Clear Architectures, Open Documentation	✓	◑	○
Data Handling	Data Privacy → Reviewing data rules → Secure Transfer	✓	◑	✕
Change Management	Controlled Updates → Reviewing data rules → Version Control	✓	◑	○
Security	Access Controls → Threat Detection	✓	◑	⊗
Contractual Protections	Clear Terms, Analyzing contracts, Liability Coverage	✓	◑	⊗

This scorecard provides a structured framework to evaluate and compare LLM vendors across critical operational and risk domains.

11.8 Operational Controls for Safe Deployment

Guardrails are the technical enforcement layer that sits between user inputs and model responses. They implement content rules, topic restrictions, output format requirements, and escalation triggers at inference time. Guardrail systems allow organizations to define policy rules in a structured format and enforce them programmatically rather than relying on the model's own judgment. The model's judgment is a control, not a guarantee; guardrails are the enforcement mechanism that backstop it.

Sandboxing restricts the capabilities available to an LLM agent in agentic deployments. A model that can execute code should execute code only in an isolated environment with no network access and no access to production data. A model that queries a database should have read-only access to the minimum tables required for its task. A model that can send communications should have its outbound communications reviewed by a human or a moderation layer before transmission. The principle of least privilege, giving the model only the capabilities it needs for its specific function, reduces the blast radius of any prompt injection or behavioral anomaly.

Capability limiting is a governance decision that must be made explicitly for each agentic LLM deployment. The temptation is to give the model broad capabilities and rely on system prompt instructions to constrain its behavior. This approach produces a system that is more capable than it needs to be and therefore more dangerous when it behaves unexpectedly. The principle of minimal capability starts with the smallest set of actions the system needs. It adds capability only when demonstrated need and demonstrated safety justify it, producing more governable agentic systems.

Diagram 11.7: Agentic LLM Deployment with Least-Privilege Sandboxing

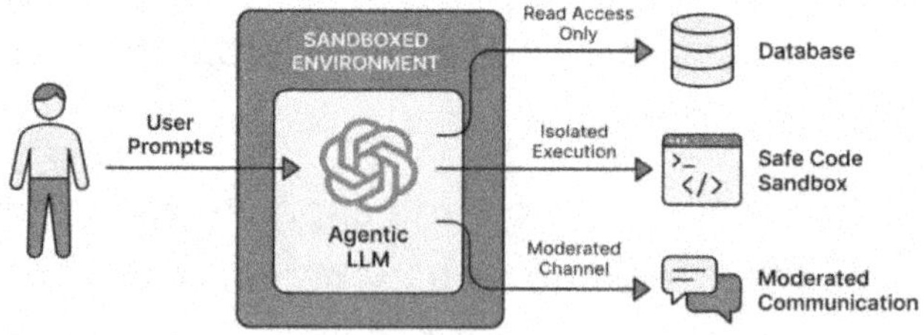

Ensures secure deployment by confining LLM actions to a sandboxed environment with controlled, read-only permissions.

11.9   Manager's Checklist

Before deploying any LLM system in a production context: Define the threat model, identify which of the four primary failure modes (hallucination, prompt injection, content safety, data disclosure) are highest risk for this use case, and what stakeholder level applies. Write and version-control the system prompt before deployment, with an approval record. Conduct red-team testing with documented methodology, scope, findings, and mitigations. Implement input- and output-moderation classifiers and test their false-positive and false-negative rates on use-case-specific test inputs. Define human verification thresholds for high-stakes output categories and design the verification workflow with qualified reviewers. Implement a context window hygiene audit to verify what enters the context window, at what sensitivity level, and under what authorization.

Vendor management: Execute a data processing addendum before sending any data to a third-party model provider. Verify that model version change notification provisions are in the contract. Establish a model version testing protocol so that provider updates are tested before being forced into production. Maintain a provenance log of each inference call with sufficient metadata to reconstruct what model version processed what input. Review vendor security certifications and incident notification commitments annually. Apply the five-domain vendor assessment scorecard to each provider at initial engagement and at renewal. Assess provider concentration risk and document mitigation plans for high-stakes deployments.

11.10 Governing Generative AI as a System

Generative AI governance is not a set of features to be checked before launch. It is an ongoing operational discipline that requires dedicated engineering, continuous monitoring, and systematic vendor accountability. The organizations that deploy LLMs most successfully are those that treat governance architecture as a first-class design requirement — not an afterthought added when something goes wrong. The controls described in this chapter are achievable with current technology and reasonable organizational effort. The question is not whether they are possible but whether the organization has built the governance muscle to implement them consistently across every LLM deployment in its portfolio.

The professional services firm implemented each of the controls described here following its three incidents. The remediation took four months and required rebuilding the integration architecture, implementing a guardrail system, and rewriting the system prompt under formal change control. The cost, in engineering and delayed deployment of subsequent use cases, was substantial. The same controls, built before the initial deployment, would have taken six weeks to complete.

Claude Louis-Charles, PhD

Governance is most expensive when it is treated as remediation rather than design.

12 AI Audit, Assurance, and Continuous Compliance

An internal audit team at a regional bank selected its AI lending model as the subject of a routine model risk review. The review team requested the model's validation report, training data documentation, fairness assessment, and records of ongoing monitoring. After two weeks of searching, the model owner produced a validation report from eighteen months prior, no training data documentation, a fairness analysis that covered two protected classes when the applicable regulation required five, and monitoring dashboards that had been disconnected when the model was migrated to a new serving infrastructure six months earlier. The model had been running in production without effective monitoring for half a year. The audit finding triggered a remediation plan, a model suspension, and a requirement to conduct retroactive impact analysis on decisions made during the gap period. The cost of not having an assurance program was several times the cost of building one.

The audit team's most uncomfortable finding was not the missing documentation. It was that the model had continued making consequential lending decisions for six months without anyone noticing that the monitoring infrastructure had gone dark. No alert had fired when the monitoring connection broke. No one had noticed the absence of monitoring reports. No escalation had occurred. The governance failure was not limited to documentation; it was a failure of operational awareness. The assurance program is the infrastructure that prevents that failure mode.

An audit-ready AI governance program is not a program built for auditors. It is a program built for the organization, one that happens to produce the evidence auditors need as a natural byproduct of good operational discipline. The distinction matters because programs built primarily for audit theater produce documents that pass review but do not reflect actual practice. Programs built for operational discipline produce accurate evidence by tracking what the organization actually did, not what it wishes it had done. Auditors and regulators are experienced at distinguishing between the two.

The regulatory and legal environment for AI is moving toward mandatory audit requirements in high-stakes sectors. The EU AI Act requires conformity assessments and ongoing technical documentation for high-risk AI systems. Federal financial regulators have issued supervisory guidance on model risk management that applies to AI systems. State insurance regulators have begun requiring algorithmic fairness assessments. Federal healthcare oversight bodies have proposed audit requirements for AI-assisted clinical decisions. Managers who wait for these requirements to become binding before building assurance programs will find themselves building under pressure, with inadequate data and no historical baseline.

This chapter constructs the assurance program that a manager can build proactively. It covers the audit trail architecture needed to make AI systems auditable, the evidence collection framework that populates the audit trail, the automated monitoring systems that maintain continuous compliance between audits, and the playbooks for responding to internal and external audit findings. The chapter also provides audit scope templates that can be used directly or adapted for specific deployment contexts.

Diagram 12.1: AI Assurance Program Architecture Overview

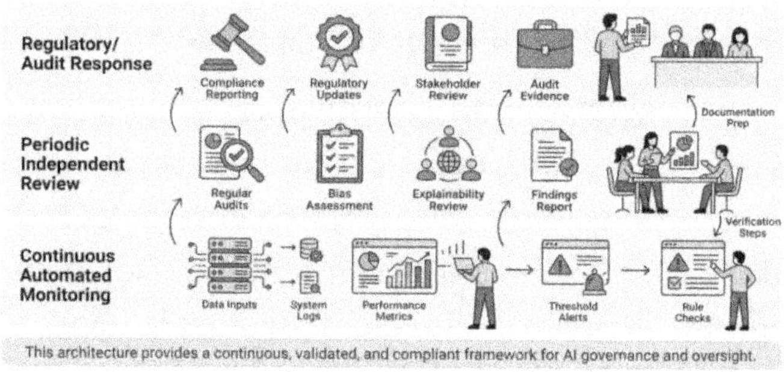

This architecture provides a continuous, validated, and compliant framework for AI governance and oversight.

12.1   The Assurance Program

An AI assurance program is the organizational framework that provides management, the board, regulators, and affected stakeholders with reliable evidence that AI systems are performing as intended, within defined risk tolerances, and in compliance with applicable requirements. The program has three components: a control framework specifying what must be in place, a monitoring architecture providing continuous visibility into control operation, and an audit function providing periodic independent validation that the controls are working. These components are not substitutes for one another; they are complementary, and each provides a different type of assurance.

The control framework is the foundation. Without defined controls, there is nothing to monitor or audit. Controls must be documented to specify the control objective, the control activity, the responsible party, the evidence they produce, and the testing methodology before monitoring and audit can be designed around them. A control that exists in practice but is not documented does not provide assurance; it provides only luck. A documented control that is not practiced is documentation fraud. The assurance program's credibility depends on the alignment between what is documented and what actually happens.

The scope of the assurance program should encompass every AI system deployed in production that produces outputs affecting individuals or business decisions. This includes models that were deployed before the governance program was established. Legacy systems present higher governance risk than new deployments, precisely because they have accumulated operational history without systematic oversight. The inventory of in-scope systems should be reviewed quarterly to ensure that new deployments are captured and retired systems are closed out appropriately.

12.1.1 Program Governance Structure

The assurance program requires a governance structure that defines who owns it, who executes it, and who provides independent oversight. A common pattern is the three-lines model: the first line (AI system owners and development teams) owns and operates the controls; the second line (risk management and compliance) designs the control framework, monitors compliance, and reports to senior management; the third line (internal audit) provides independent assurance to the board and audit committee that the controls are effective. Each line has defined responsibilities, reporting lines, and cadences for delivering its assurance outputs. Without this structure, assurance activities tend to be duplicative in some areas and absent in others.

The independence of the third-line function is non-negotiable. Internal audit must have direct access to AI systems, their documentation, and their operational records without going through the system owners. If the internal audit must request access through the model development team, the team can filter what is available, thereby compromising the review's independence. Access controls and audit rights must be established when systems are deployed, not negotiated on an audit-by-audit basis.

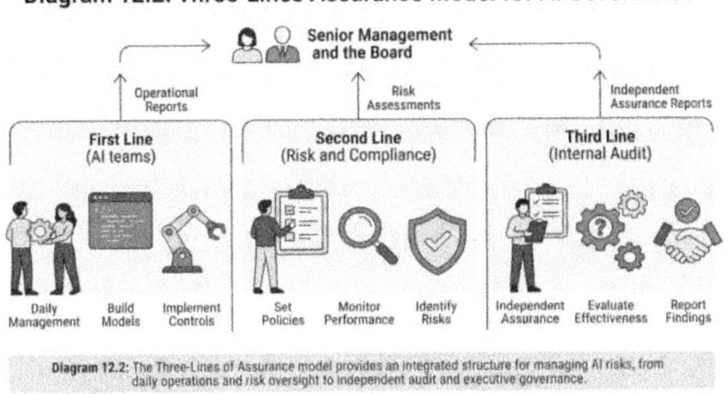

Diagram 12.2: The Three-Lines of Assurance model provides an integrated structure for managing AI risks, from daily operations and risk oversight to independent audit and executive governance.

12.2   Model Audit Trails

A model audit trail is the complete, chronological record of everything that happened to a model from its creation to its retirement: who created it, on what data, with what methodology, with what validation results, under what deployment conditions, with what performance in production, subject to what changes over time, and ultimately retired or replaced under what circumstances. This record must be immutable, meaning it cannot be modified retroactively and must be tied to specific model versions so that the audit trail for version 1.0 remains accurate even after version 2.0 is deployed.

The technical implementation of an immutable audit trail varies by infrastructure. Still, the logical requirements are consistent: append-only logging, cryptographic hashing of log entries to detect tampering, access controls that prevent the model owner from modifying the trail, and retention policies that keep the trail for at least the statute of limitations applicable to the use case. For financial services, this is typically five to seven years. For healthcare, it aligns with medical record retention requirements. For employment decisions, the EEOC recommends a minimum of four years for automated selection criteria.

The granularity of the audit trail should be calibrated to risk. For high-stakes systems, the trail should capture individual inference records, including the input presented, the output produced, the model version that processed it, and the applicable timestamp. For medium-stakes systems, inference-level logging may not be required, but the trail should capture aggregate performance metrics, model changes, and material events. For low-stakes systems, a change log and periodic performance summaries may be sufficient. The granularity decision should be documented in the system's governance plan and reviewed when the system's risk classification changes.

12.2.1 Model Card as Audit Anchor

The model card, a structured document describing a model's purpose, capabilities, limitations, training data, evaluation results, and intended uses, serves as the human-readable anchor for the audit trail. Each model version should have its own model card, version-controlled alongside the model artifact. The model card does not replace the detailed technical audit trail. Still, it provides the interpretive context that makes the technical record meaningful to auditors, compliance officers, and non-technical reviewers. A well-maintained model card substantially reduces the time and cost of both internal reviews and regulatory examinations.

Model cards should be written to be understood by the audit audience, not by the development team. This means avoiding model-specific jargon, expressing performance metrics in terms of decision impact rather than technical statistics, and being explicit about limitations in plain language. A model card that describes a 2% error rate without contextualizing what kinds of errors occur, how frequently they affect specific populations, and what the impact of an error is for the affected individual, has provided technical information without governance insight.

12.3   Evidence Collection

Evidence collection is the systematic process of assembling and preserving the artifacts that demonstrate control operation. Unlike retrospective documentation, assembling evidence after an audit request has been received, prospective evidence collection captures evidence as a natural byproduct of the governance process. The difference is consequential: retrospective evidence can be incomplete, inconsistent, or inadvertently reconstructed; prospective evidence is contemporaneous and therefore more reliable and more credible.

The evidence collection framework maps each control to the artifacts it produces, specifies the format and storage location for each artifact, and defines the retention period and access controls. For each AI system, the evidence portfolio should include: training data documentation (source, composition, preprocessing steps, known limitations), validation reports (methodology, test results, identified weaknesses), deployment approval records (who approved, what conditions were attached), monitoring outputs (periodic reports, threshold alerts, anomaly investigations), change records (what changed, when, who approved, what re-validation was performed), and incident records (what happened, what was the impact, how was it resolved).

The evidence framework should specify not only what artifacts are required but what "adequate" means for each artifact type. A validation report is adequate if it covers the required test categories, was conducted by a party with defined independence from the development team, documents the test methodology with enough specificity to be reproduced, and records the findings and their dispositions. A validation report that is a summary slide deck prepared after the fact does not meet the adequacy standard, even if it technically contains the required categories of information. The adequacy criteria are the operational definition of "evidence," and they must be specified clearly enough that the people producing the artifacts know whether they are meeting the standard.

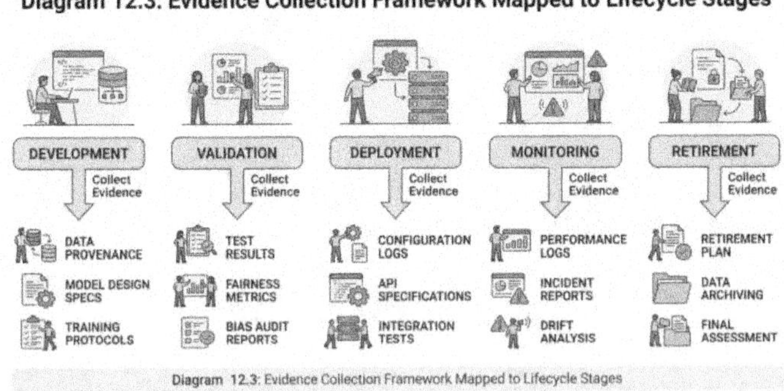

Diagram 12.3: Evidence Collection Framework Mapped to Lifecycle Stages

Diagram 12.3: Evidence Collection Framework Mapped to Lifecycle Stages

12.3.1 Evidence Integrity Controls

Evidence that can be modified or selectively deleted is not reliable. The integrity of the evidence collection depends on controls that prevent unauthorized modification and ensure completeness. Minimum integrity controls include: storing evidence in a system where the model owner does not have delete or modify permissions, timestamping all artifacts at the time of creation using a trusted time source, maintaining an evidence inventory that lists every required artifact for each model with its storage location and last-verified date, and conducting quarterly reviews to confirm that required artifacts are present and have not been modified.

The evidence inventory is the governance tool that makes the evidence portfolio manageable at scale. For each model in the portfolio, the inventory lists the required artifacts, their storage locations, their version numbers, their last-review dates, and any identified gaps. The inventory is reviewed in the monthly governance meeting and updated whenever a model is added, changed, or retired. A model without complete inventory entries is a model without complete governance coverage, and that gap should generate a risk register entry until it is resolved.

12.4 Automated Monitoring

Automated monitoring translates the organization's risk tolerance and compliance requirements into continuous, machine-executed checks that operate without human attention. A well-designed monitoring system provides early warning of performance degradation, discriminatory drift, data quality issues, and anomalous usage patterns, often detecting problems weeks before they become visible in downstream outcomes. The investment in monitoring infrastructure pays for itself in prevented incidents and reduced audit remediation costs.

The monitoring architecture for AI systems has four layers. Data quality monitoring checks the inputs to the model: are they arriving in expected formats, within expected value ranges, and without distribution shifts that would invalidate the model's assumptions? Model performance monitoring checks prediction accuracy, confidence calibration, and output distributions against baselines established at deployment. Fairness monitoring checks disaggregated performance metrics across protected groups against baseline fairness measurements. Operational monitoring checks infrastructure health: latency, error rates, throughput, and availability. Each layer has different cadence requirements; data and operational monitoring typically run continuously, while fairness monitoring may be computed daily or weekly.

Monitoring alerts require defined response SLAs. An alert that fires but generates no response within a defined period has not served its governance function. For critical alerts indicating a control failure with the potential for immediate harm, the response SLA should be measured in hours. For significant alerts, performance degradation, or fairness drift approaching the threshold, the SLA may be one to three business days. For informational alerts — trend signals that have not yet crossed thresholds, a weekly review may be appropriate. The response SLAs and the escalation path for unresponded alerts must be documented and tested before the monitoring system goes live.

12.4.1 Drift Detection

Model drift occurs when the statistical relationship between inputs and outputs changes over time, typically because the world has changed in ways the training data did not anticipate. There are two primary drift types: data drift, where the input distribution changes, and concept drift, where the underlying relationship the model learned no longer holds. Both can degrade model performance without triggering obvious errors. Automated drift detection compares the current input distribution to the training distribution using statistical tests (Kolmogorov-Smirnov, population stability index) and alerts when drift exceeds defined thresholds. Drift alerts trigger a review cycle: evaluate whether performance has degraded, determine whether re-training is required, and document the decision and its rationale.

Not every detected drift requires retraining the model. The drift review cycle evaluates whether the drift is material, whether it has degraded performance to the point of concern, or whether it represents a demographic shift that changes fairness assessments. If the drift is material and affects a high-stakes system, retraining or model replacement is required. If the drift is minor and performance remains within acceptable bounds, the drift record and the decision not to retrain should be documented in the model's evidence portfolio. This documentation demonstrates that the organization detected the drift and exercised informed judgment about its significance.

Continuous multi-stage monitoring ensures system reliability, data integrity, and ethical model deployment.

12.5 Independent Reviews

Continuous monitoring provides real-time visibility but does not substitute for periodic independent review. An independent review ensures that the monitoring itself is functioning correctly, that the control framework is being implemented as designed, and that no systematic governance gaps have developed over time. The "independent" in independent review means that the reviewers have no direct stake in the model's continued deployment and sufficient technical competence to evaluate the controls they are reviewing. Independence without competence is a rubber stamp. Competence without independence is a conflict of interest.

The cadence and scope of independent reviews should be calibrated to risk. High-stakes AI systems that affect credit, employment, healthcare, benefits, or public safety should receive a comprehensive independent review at least annually. The review should cover: the currency and accuracy of the model card, the completeness of the evidence portfolio, the adequacy of the monitoring architecture, the fairness of model outcomes in the review period, the appropriateness of the human oversight mechanisms, and the quality of incident responses during the period. Medium-stakes systems may receive lighter-touch reviews on a longer cycle. Low-stakes systems may be reviewed through a portfolio-level sampling approach rather than individually.

Independent reviews should produce written reports with findings classified by severity, management responses, and remediation timelines. The findings, management responses, and remediation status should be maintained in a searchable repository that allows tracking across multiple review cycles. A finding that appears in multiple successive reviews without remediation is a governance failure that requires senior management attention, not just another cycle of the standard response process.

Diagram 12.5: Annual Independent Review Scope and Process

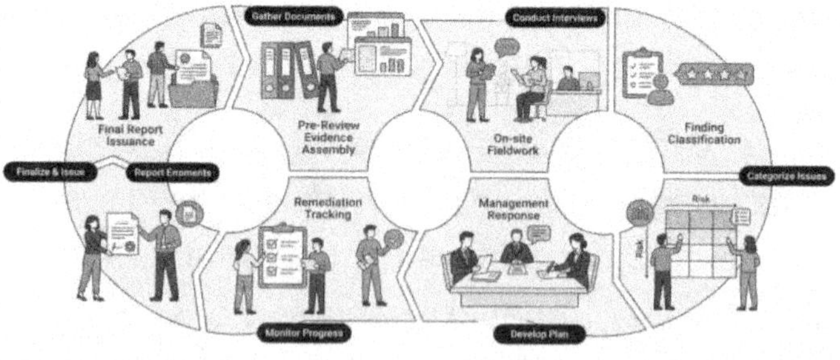

This infographic outlines the standardized steps for the annual independent review, ensuring comprehensive assessment and timely resolution of findings.

12.6   Audit Scope Templates

Standardized audit scope templates reduce the setup cost for each review and ensure consistency across reviews of different systems. A scope template specifies the review objectives, the control domains in scope, the evidence items to be requested, the testing procedures for each control, and the finding classification criteria. The template is a starting point that is adapted for each engagement based on the system's risk profile, deployment context, and applicable regulatory requirements.

The core control domains for an AI system audit scope include: model development and validation controls, data governance and provenance, fairness and discrimination risk management, transparency and explainability, human oversight mechanisms, incident management, change management, and vendor management for externally provided model components. Each domain is associated with a standard set of evidence requests and testing procedures. The testing procedures specify not just what to check but how to check it: which documents to review, which samples to test, which calculations to perform independently, and what constitutes adequate evidence of control operation.

Scope templates should be version-controlled and updated annually to incorporate new regulatory requirements, lessons learned from completed reviews, and emerging governance standards. The version history of the template is governance evidence in itself: it demonstrates that the organization has maintained its review methodology over time and has updated it in response to the evolving AI governance environment. Templates that have not been updated in several years are a governance signal that the organization's assurance methodology is not keeping pace with the risk environment.

12.6.1 Finding Classification

Audit findings should be classified by severity to support prioritized remediation. A standard three-tier classification works for most AI audit programs: critical (a control is absent or non-functional and immediate action is required to prevent material harm), significant (a control is partially implemented or its evidence is inadequate, requiring remediation within a defined period), and observation (a control is functional but could be strengthened, with remediation recommended but not mandated). Critical findings should trigger immediate escalation to senior management and, where required by regulation, notification to the applicable regulator. Each finding should include a specific description, the root cause, the potential impact, and the recommended remediation.

Finding language matters. A finding that says "monitoring is inadequate" does not tell the model owner what needs to change. A finding that says "the fairness monitoring system was disconnected from the production serving infrastructure on [date] and has not produced reports since; the absence of monitoring output generated no alerts; no review occurred during the six-month gap" gives the model owner a specific, actionable description of what happened and why it is a problem. Precise findings lead to specific remediations; vague findings lead to vague responses that are difficult to verify as complete.

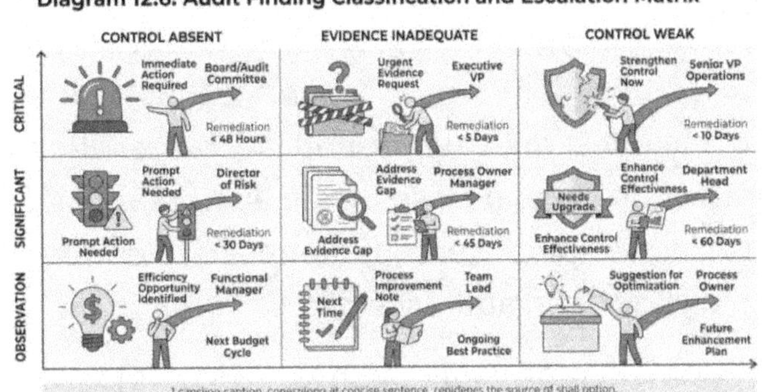

Diagram 12.6: Audit Finding Classification and Escalation Matrix

12.7 Responding to Findings

Responding well to audit findings is as important as having strong controls in the first place. Regulators and auditors evaluate both the control environment and the organization's capacity to identify and correct problems. A finding that is acknowledged, root-caused, and remediated systematically demonstrates governance maturity. A finding that is disputed without substance, remediated superficially, or recurs in subsequent reviews demonstrates the opposite. The response to findings is itself a governance signal.

The standard response process for audit findings follows four steps: management response, root cause analysis, remediation planning, and remediation verification. Management response acknowledges the finding, either agrees with or contests it with a specific rationale, and commits to a remediation timeline. Root cause analysis identifies the underlying reason the control failed, whether it was a design gap, a resource gap, a training gap, or a process failure. The remediation plan addresses the root cause, not just the symptom, and includes milestones and an owner. Remediation verification is conducted by the audit team or a second-line function on the completion date to confirm that the finding has been addressed.

Remediation verification is the step most commonly skipped under time pressure, and skipping it is consistently consequential. Without verification, the assurance function has no evidence that the remediation was completed as described. Management responses saying "we have implemented the recommended control" are not evidence of implementation; they are assertions. The verification step converts assertions into evidence by testing whether the control is actually in place and functioning. Organizations that skip verification often rediscover the same findings in the next review cycle.

12.7.1 Regulatory Inquiry Response

When a regulatory inquiry relates to AI systems, the governance program's evidence portfolio is the organization's primary response asset. The response process should begin with a rapid evidence inventory: identify every AI system potentially in scope for the inquiry, locate each system's current evidence portfolio, and identify gaps that must be addressed or disclosed. Legal counsel should be engaged from the outset to guide the response strategy, manage privilege determinations, and ensure that the response is complete and accurate. A response that later proves incomplete or misleading, even unintentionally, creates substantially greater regulatory exposure than the original inquiry.

The inquiry response should not reconstruct evidence retroactively. Regulators are experienced at identifying documents whose metadata suggests they were created after the inquiry was received. Retrospective documentation is both legally risky and practically counterproductive: it suggests that the governance program exists primarily on paper and that the organization is building its compliance story in response to scrutiny rather than as a matter of operational discipline. The evidence assembled in response to a well-designed governance program should be contemporaneous; the response process should retrieve it rather than create it.

Diagram 12.7: Regulatory Inquiry Response Playbook

This structured playbook ensures timely, accurate, and compliant responses to regulatory inquiries.

12.8   Manager's Checklist

Assurance program foundations: Define the three-line accountability structure for AI assurance and document the responsibilities of each line. Establish an immutable audit trail for each production AI system with appropriate granularity for its risk classification. Maintain version-controlled model cards for every deployed model, reviewed annually, and updated on each model change. Build an evidence-collection framework that captures artifacts contemporaneously throughout the AI lifecycle and defines adequacy standards. Implement automated monitoring of data quality, model performance, fairness metrics, and operational health, with defined alert thresholds and response SLAs, schedule and resource annual independent reviews for all high-stakes AI systems, and track written reports and findings.

Ongoing operations: Conduct quarterly evidence inventory reviews to confirm required artifacts are present, adequately documented, and unmodified. Run drift detection checks continuously and review alerts within the defined response SLA document for every alert and its disposition. Document all model changes, including re-validation status and approval records, before deploying the change. Maintain an open finding log that tracks all audit findings to closure, with management responses, remediation plans, and verification records. Test the regulatory inquiry response process annually through a tabletop exercise covering evidence assembly, legal privilege, and escalation procedures. Update the audit scope template annually to reflect new regulatory requirements or changes to the governance framework.

12.9 Assurance as Operational Discipline

An effective AI assurance program is not a documentation exercise conducted before an audit and shelved afterward. It is the continuous operational discipline that makes the organization's AI portfolio governable at scale. The monitoring architecture, the evidence collection framework, and the audit function work together to provide the organization, not just the auditor, with reliable information about how its AI systems are performing and whether its governance commitments are being honored. Managers who build this discipline proactively will find that audits become confirmation events rather than discovery events, and that regulatory inquiries become manageable exercises rather than crises.

The investment required is real but bounded. The audit trail, monitoring, and evidence frameworks described here are achievable with standard enterprise tooling. The harder work is cultural: building the organizational habit of contemporaneously capturing evidence, treating monitoring alerts as decision triggers rather than noise, and maintaining the independence of the audit function against the operational pressure to use it as a rubber stamp. That cultural work is a governance challenge, not a technical one, and it starts with the manager's own expectations about what assurance means.

The bank from the opening scenario completed its remediation retroactive impact analysis, new monitoring infrastructure, and updated documentation. The remediation took eight months and identified several hundred lending decisions made during the monitoring gap that required manual review. The cost was measured in staff time, delayed product initiatives, and regulatory relationship capital. The assurance program that would have prevented the incident, built before deployment, would have cost a fraction of the remediation. Assurance is not overhead. It is risk capital deployed in advance, where it does the most good.

13 Operational Playbooks and Conclusion

After attending a three-day AI governance training program, a newly appointed AI risk manager returned to her organization with a clear conceptual framework and no executable plan. She understood the principles. She could diagram the risk categories, describe the five Blueprint principles, and explain the three-lines model. What she lacked was the set of ready-to-use artifacts that would let her start Monday morning: a product charter her team could fill out before the next model deployment, a risk register template her analysts could populate, a meeting agenda that would make the weekly governance cadence productive rather than performative. This chapter provides those artifacts. Principles without tools are education. Principles with tools are a program.

When she arrived Monday morning, a development team was waiting with a model ready for deployment review. They needed a governance sign-off. She had no sign-off process, no review criteria, and no artifacts to guide the conversation. She improvised, asked questions, took notes, sent an email summary, and the model was deployed that afternoon. Three months later, a monitoring alert that the team had not implemented correctly failed to fire, and a data quality issue propagated into production decisions for two weeks before a downstream analyst noticed unusual output patterns. The training had not given her the tools to prevent that sequence of events. This chapter does.

The gap between governance principles and governance execution is where AI risk programs most commonly fail. Organizations invest in frameworks, policies, and training, then discover that the people responsible for execution are uncertain what exactly to do on any given Tuesday. A model is ready for review. What does the review process look like step by step? An incident has occurred. What is the first action to take, and who owns it? The monthly governance meeting is tomorrow — what should be on the agenda and what decisions need to be made? Without answers to these operational questions, governance remains aspirational rather than actual.

This chapter is organized as a set of operational playbooks, structured, reusable artifacts that answer the "what do I do" questions at each governance touchpoint. Each playbook is designed to be adapted to the organization's specific context rather than adopted verbatim. The chapter closes with a 90-day implementation roadmap that provides a sequenced starting point for managers who are building or significantly upgrading their AI governance program.

The playbooks in this chapter distill the frameworks, controls, and processes described in the preceding twelve chapters. Each playbook is cross-referenced to the relevant chapter content so that managers who want to deepen their understanding of any component have a clear path back to the underlying framework. The goal of the playbook format is not to replace the deeper understanding that governance judgment requires, but to make common execution tasks fast and reliable, freeing the manager's attention for decisions that require substantive analysis.

Diagram 13.1: Governance Playbook Ecosystem

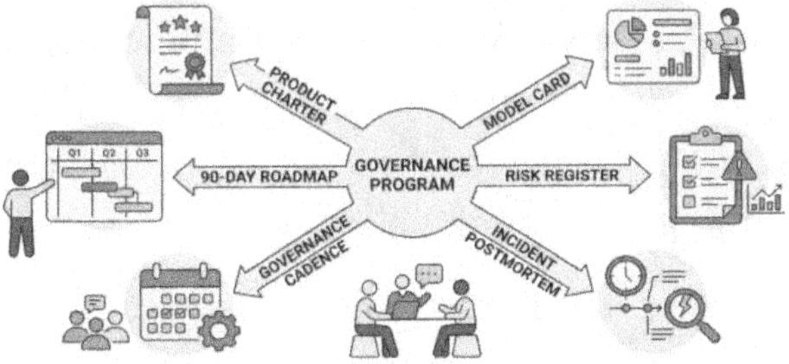

A holistic framework of interrelated artifacts guiding responsible and compliant AI development.

13.1   The Governance Playbook

A governance playbook is a structured set of decision procedures, artifact templates, and meeting protocols that operationalize the governance program. It answers the procedural questions that policies and frameworks deliberately leave open: not just "fairness assessments must be conducted" but "who conducts them, using what methodology, at what lifecycle stage, and what format does the result take." Policies define requirements. Playbooks specify execution.

The six playbook components in this chapter are designed to cover the most frequent governance touchpoints: before a new model is deployed (product charter), at model completion (model card), as risk is identified and tracked (risk register), when something goes wrong (incident postmortem), in recurring management meetings (governance cadence), and when building the program from scratch (90-day roadmap). Together, they cover the full governance lifecycle. Managers should review each template, adapt the fields to their organization's terminology and data systems, assign owners to each artifact, and build the artifact production into their existing workflows rather than treating it as additional overhead.

Playbook adoption is most successful when it is embedded in existing processes rather than added on top. If the organization already has a software deployment gate, a step where a system must be approved before it can move to production, the product charter review should be attached to that gate. If there is already a monthly technology risk meeting, the AI risk register review should be added to its agenda. Building governance artifacts into existing workflows reduces the perceived overhead and ensures that governance is not the first thing to be skipped when the team is under deadline pressure.

13.2   Product Charter Template

The AI product charter is the governance artifact that must exist before development begins on any AI system. It documents the system's intended purpose, the business problem it solves, the population it affects, the decisions it supports or makes, the data it will use, the success criteria, the risk profile, and the governance requirements that apply given that risk profile. The charter is the first artifact in the model's audit trail and the document to which all subsequent governance artifacts refer.

The charter template includes the following required sections. System identification: system name, owner, sponsoring business unit, and development team. Purpose and scope: the business problem being addressed, the decision or workflow the system will support, the intended users, and the population of individuals whose data will be processed or who will be subject to decisions. Risk classification: the stakes level (high, medium, low) based on the decision domain and potential harms, with the reasoning documented. Data authorization: confirmation that data sources have been reviewed for lawful basis, consent adequacy, and data minimization compliance. Governance requirements: the specific controls required given the risk classification — validation methodology, fairness assessment scope, human oversight mechanism, explanation requirement, and monitoring protocol.

The charter should be brief, two to four pages, and written in plain language. It is not a technical specification; it is a governance instrument read by business owners, risk managers, and legal counsel as well as technical teams. The precision required is governance precision: clear answers to governance questions, not technical detail about model architecture. A charter that runs to twenty pages because it contains technical documentation has confused two different artifacts. The technical documentation belongs in the model card.

13.2.1 Charter Approval Process

The product charter requires approval before development begins. The approval process involves three sign-offs: the system owner confirming accuracy of the purpose and scope descriptions, the risk or compliance function confirming the risk classification and governance requirements, and legal counsel confirming the data authorization review for systems that process personal data. High-stakes systems additionally require approval from a senior risk committee or designated executive. The approval record, including the approvers, the date, and any conditions attached to approval, is stored with the charter as the first entry in the model's evidence portfolio.

Charter conditions are a valuable governance tool that is underused in practice. When the approval process identifies a concern, a data authorization question that requires follow-up, a fairness assessment scope that needs to be expanded, or a human oversight mechanism that is not yet fully designed, the approval can be granted with a condition rather than deferred. The condition specifies what must be completed before deployment, not before development. This allows development to begin while the governance question is resolved, without allowing deployment to proceed until it is resolved. Tracking and closing conditions are part of the second-line risk function's role.

Diagram 13.2: Product Charter Approval Workflow

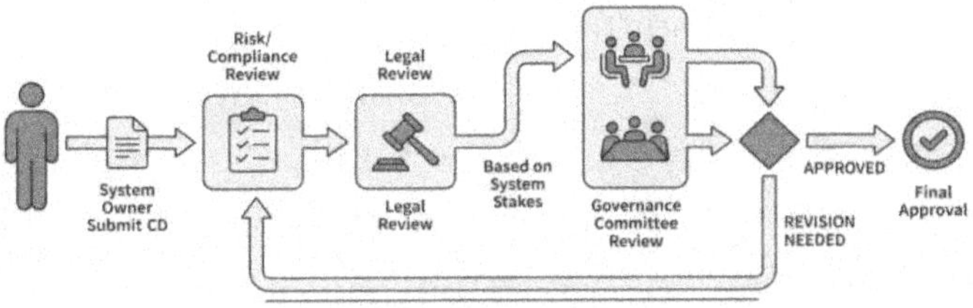

This workflow ensures thorough review and governance compliance for all new product charters before implementation.

13.3   Model Card Framework

The model card is a structured technical document that describes a completed AI model: what it does, how it was built, how it performs, and its limitations. Model cards were formalized in a 2019 paper by Mitchell et al. and have since been adopted by major AI providers and required by several regulatory frameworks as a documentation standard. For enterprise deployment, the model card serves as the primary reference document for auditors, reviewers, and oversight personnel who need to understand a system without access to the underlying code.

The enterprise model card framework includes the following required sections. Model overview: name, version, type, intended use, and out-of-scope uses. Development details: training data sources with composition statistics, preprocessing methodology, and known data quality issues. Performance metrics: overall accuracy or primary metric, disaggregated performance by relevant demographic groups, confidence calibration, and performance on defined stress tests. Fairness evaluation: the fairness assessment methodology, the groups evaluated, the metrics used, the results, and any residual disparity that could not be eliminated, with the rationale for accepting it. Limitations and risks: known failure modes, distributional boundaries, adversarial vulnerability, and recommended mitigations. Deployment requirements: the human oversight mechanism, the monitoring protocol, the re-validation trigger conditions, and the intended retirement criteria.

The model card's limitations section deserves particular care. It is the section most likely to be written defensively with vague language that acknowledges limitations in principle without specifying them. A limitations section that says "model performance may vary in edge cases" provides no useful information. A limitations section that says "model accuracy falls below the acceptable threshold for applicants who have been primary address holders for less than six months, representing approximately 3% of the applicant population; these applications are automatically routed to manual review" provides the specific operational information that oversight personnel need. Specificity in the limitations section is a governance commitment, not a liability.

13.3.1 Keeping Model Cards Current

A model card is only useful if it is up to date. Each time a model is updated with new training data, a changed architecture, or a modified serving environment, the model card must be updated to reflect the new version. Version updates should be tracked in a changelog section of the model card so that reviewers can see the model's evolution. The model card owner should review the card at least annually, even if no technical changes have been made, to confirm that the described deployment context, intended uses, and known limitations still accurately reflect the system in operation.

The most common model card failure mode in enterprise settings is not omission; it is staleness. Organizations create model cards at deployment and never update them. After two years of model updates, new data, changed business context, and evolving regulatory requirements, the model card describes a different model in a different environment. An audit that relies on a stale model card will reach conclusions about the old model rather than the current one. Model card update obligations should be part of the change management process. Whenever a change record is opened for a model, the model card update should be a required deliverable before the change is closed.

13.4   Risk Register Template

The AI risk register is the living document that tracks identified risks across the AI portfolio. Unlike a one-time risk assessment, the risk register is updated continuously as new risks are identified, existing risks evolve, and mitigations are implemented. It provides management and the second-line risk function with a current view of the AI portfolio's residual risk posture and a record of how identified risks have been addressed over time.

The risk register template has a row for each identified risk and columns for: risk identifier (unique, permanent), risk description (specific and concrete not "model may fail" but "model performance may degrade for applicants aged 60+ based on observed training data gap"), risk category (bias, accuracy, data, security, operational, compliance, vendor), likelihood rating (high, medium, low with defined criteria), impact rating (high, medium, low with defined criteria), inherent risk rating (product of likelihood and impact), current mitigating controls, residual risk rating after controls, risk owner, and review date. An additional comments field captures reasoning that does not fit the structured columns.

Risk descriptions require effort to write well. The temptation is to write descriptions that are general enough to be safe, vague risks cannot be definitively shown to have materialized. But a risk register populated with vague risks is not a governance tool; it is a checklist that creates the appearance of risk management without its substance. The test of a well-written risk description is whether a manager who did not write it can read it, understand specifically what could go wrong, and assess whether the mitigating controls described address the risk. If that test fails, the description needs revision.

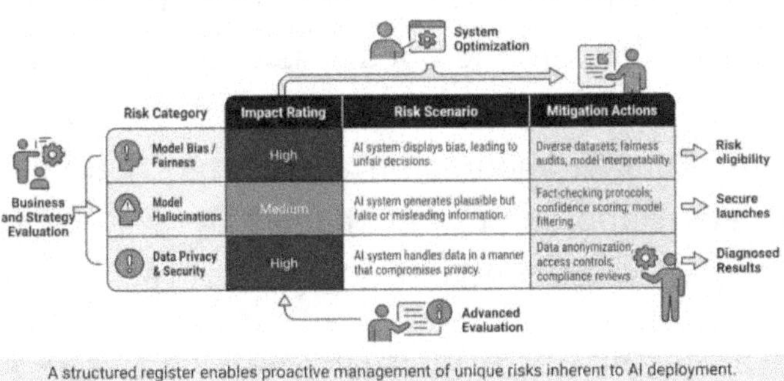

Diagram 13.3: AI Risk Register Structure and Sample Entries

A structured register enables proactive management of unique risks inherent to AI deployment.

13.4.1 Risk Register Cadence

The risk register for the AI portfolio is reviewed in the monthly governance meeting (see Section 13.6). Each risk owner is responsible for updating their risk entries before the meeting and flagging any changes to likelihood, impact, or mitigation status. New risks identified since the last meeting are added and assigned to owners at the meeting. Risks that have been fully mitigated are closed with a closure date and disposition note, and retained in the register history rather than deleted. The closed risks provide a historical record of the organization's risk management activity that is useful in both internal and external reviews.

Risk register reviews should not degenerate into status updates. The governance value of the review lies in discussing whether the risk picture has changed, whether mitigations are effective, and whether new risks have emerged that require attention. A review that consists entirely of "risk X is still open, owner Y is working on it" is an administrative exercise, not a governance exercise. The meeting facilitator should drive the discussion toward the questions that require judgment: Is this residual risk level still acceptable? Have circumstances changed in a way that requires a new mitigation approach? Are there emerging risks in the external environment that we have not yet captured?

13.5   Incident Postmortem Template

AI incidents, events in which a model produces harmful, erroneous, or non-compliant outcomes, require structured analysis to prevent recurrence. The incident postmortem is the governance artifact that ensures this analysis happens systematically rather than informally. A well-conducted postmortem produces actionable findings: root causes that can be addressed, control gaps that can be closed, and process changes that reduce the likelihood of recurrence. An incident postmortem that concludes only that "the model made an error" has not done the analytical work.

The incident postmortem template includes: incident summary (brief description, detection date, resolution date, systems involved, and population affected), timeline (chronological sequence of events from first indicator to full resolution, with timestamps and responsible parties), impact assessment (harm caused, number of affected individuals, business impact, regulatory implications), root cause analysis (five-whys or equivalent structured analysis tracing the incident to its underlying causes), contributing factors (organizational, technical, or process conditions that made the incident more likely), remediation actions (specific changes implemented or planned, with owners and completion dates), and governance lessons (changes to policies, controls, monitoring, or training triggered by this incident).

The impact assessment section requires careful handling for incidents involving personal data or individual harm. The organization must assess not only the business impact but also the impact on affected individuals: which decisions were affected, how many individuals were affected, the nature of the potential harm, and which remediation or notification obligations apply. For incidents that may have produced discriminatory outcomes, the impact assessment must include a demographic analysis of who was affected. This analysis feeds both the remediation plan and any required regulatory notification.

13.5.1 Blameless Analysis

Effective postmortems are blameless in their analytical posture. The goal is to understand how the system's technical and organizational aspects led to the incident, not to assign personal fault to individuals operating within the systems and processes available to them. This posture is not about avoiding accountability; accountability is established through the risk register, the control framework, and the remediation plan. It is about creating the psychological safety needed for people to report incidents accurately and completely, which is a prerequisite for learning. Organizations with blame-oriented incident cultures systematically underreport, which means they systematically under-learn.

The governance lesson is a required section of every postmortem, not an optional reflection. Incidents that do not produce governance changes teach the organization nothing, and the failure to learn is itself a governance finding. Governance lessons should be specific: not "we need better monitoring" but "the monitoring alert for data quality anomalies was configured with a threshold that was too high to detect the specific data shift that produced this incident; the threshold will be lowered from X to Y and tested against the incident's input data before the change is deployed." Specific lessons produce specific changes; general lessons produce general intentions.

Diagram 13.4: Incident Postmortem Workflow from Detection to Closure

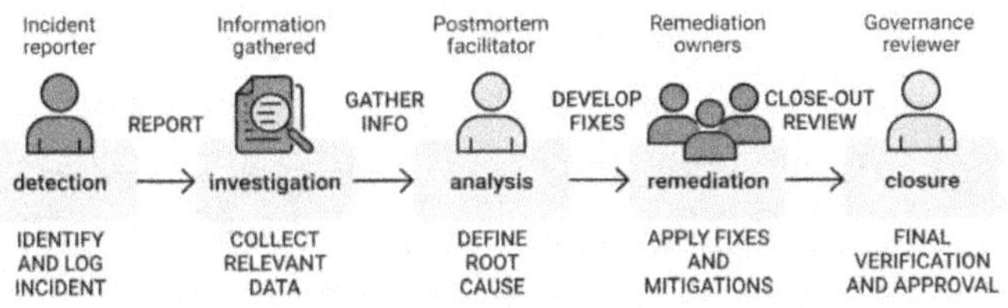

The incident postmortem system provides final approval to the incident, ensuring it is officially closed and reviewed.

13.6   Governance Cadence

A governance cadence is the set of recurring meetings and reporting cycles that keep the AI governance program operational between crises. Without a defined cadence, governance activity tends to be event-driven, reactive to incidents, audits, and regulatory inquiries rather than proactive. A defined cadence creates the organizational habits that transform governance from a set of documents into a living program.

The recommended cadence structure has three tiers. Weekly: a brief operational standup (30 minutes) covering monitoring alerts from the past week, open incidents, and deployment activities in progress.

Attendees: AI system owners, second-line risk representative, and the AI governance lead. Monthly: a governance review (90 minutes) covering the risk register, new model deployment approvals, audit findings and remediation status, and policy or regulatory updates. Attendees: all AI system owners, risk and compliance, legal, and senior management representatives. Quarterly: a strategic review (half day) covering portfolio risk trends, program maturity assessment, vendor management reviews, and governance program updates—attendees: senior management, internal audit, and external advisors as appropriate.

The reporting outputs from each cadence tier feed the next tier up. The weekly standup produces an operational status summary that informs the monthly meeting. The monthly meeting produces a risk and compliance report that informs the quarterly review. The quarterly review produces a board-level summary that provides senior leadership with a current view of the AI risk posture. This reporting chain is not bureaucratic overhead; it is the information infrastructure that allows governance decisions to be made at the appropriate level with appropriate evidence.

13.6.1 Meeting Discipline

Governance meetings are only valuable when they result in decisions and actions. Each meeting should have a pre-circulated agenda, a designated facilitator, and a documentation owner who captures decisions and actions with named owners and due dates. Actions from the previous meeting are reviewed at the start of the next meeting. Meetings that consistently fail to produce decisions because the right people are not present, because materials are not prepared, or because the agenda is too broad should be restructured rather than perpetuated. The cadence exists to maintain governance discipline; if it is not doing that, the cadence itself needs to be governed.

Attendance discipline is a governance signal. When system owners consistently miss the governance cadence, it indicates that governance is not perceived as a priority. This perception is the manager's responsibility to correct both by setting expectations and by making the meetings genuinely useful. A meeting that consists entirely of status reports with no decision-making adds no value to participants who could have read the same information in an email. Effective governance meetings create decision rights, surface trade-offs, and produce commitments that have accountability. That is what makes attendance a reasonable expectation.

Diagram 13.5: Three-Tier Governance Cadence Calendar

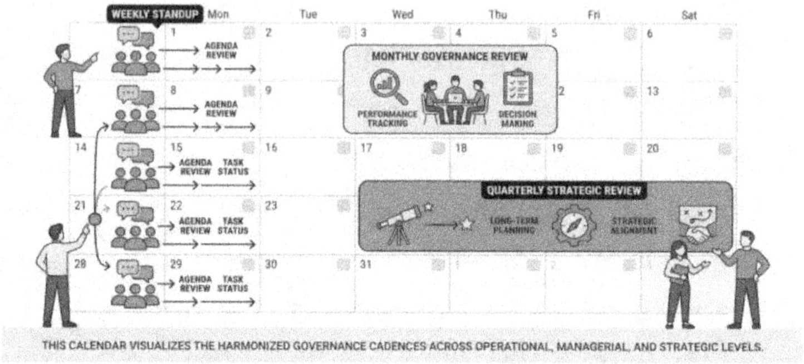

13.7   The 90-Day Implementation Roadmap

The 90-day roadmap is the execution plan for managers who are building an AI governance program from a limited baseline. It sequences the foundational work into three 30-day phases, each building on the previous, so that the program reaches a functional baseline by day 90. The roadmap is not exhaustive. A mature program will take considerably longer to build fully. Still, it defines the minimum viable governance posture that should be in place before new high-stakes AI systems are deployed.

Days 1 through 30 — Foundation: Conduct an inventory of all AI systems currently in production or in late-stage development. For each system, document the system owner, the business purpose, the population affected, and the current state of governance documentation (what exists, what is missing). Classify each system by risk level using a documented classification methodology. Assign a governance lead for the program with a defined charter. Establish the evidence repository as the storage system for maintaining governance artifacts with appropriate access controls and retention settings. Adapt the product charter, model card, and risk register templates to organizational terminology. Schedule the first monthly governance meeting and communicate expectations for attendance.

Days 31 through 60 — Documentation: For each high-stakes system, complete or update the model card and populate the risk register. For systems without monitoring, implement minimum viable monitoring covering model performance and basic data quality. For systems without a human review process, design and document one. Conduct the first monthly governance meeting and review open risks in the risk register with system owners. Identify the two or three highest-priority remediation actions and assign them to owners with defined timelines. Engage legal counsel to review data processing practices for each high-stakes system that processes personal data, and document the outcome of the authorization review. Circulate the incident postmortem template and establish the reporting pathway for incidents.

Days 61 through 90 — Assurance: Complete the evidence portfolio for each high-stakes system, addressing the most critical documentation gaps identified during the documentation phase. Implement fairness monitoring for all high-stakes systems and define alert thresholds with documented rationale. Draft the regulatory inquiry response playbook and conduct a tabletop exercise with IT, Legal, and Compliance participants. Submit a program status report to senior management covering the portfolio risk inventory, documentation status, monitoring coverage, open risks, and a prioritized list of remaining gaps with timelines. Present the governance cadence schedule to all system owners and confirm participation commitments. By day 90, the organization should have a current inventory, documented risks, functioning monitoring of high-stakes systems, a governance meeting rhythm, and a clear, committed view of what still needs to be built.

Diagram 13.6: 90-Day Implementation Roadmap with Milestone Markers

This roadmap outlines the three-stage execution for key operational readiness workstreams.

13.8   The Governance Imperative

This book has worked from premise to practice: from the societal and ethical context of AI governance, through the legal and regulatory landscape, into the operational structures, lifecycle controls, and assurance programs that make governance real. Each chapter has been built around the same conviction: governance is not compliance theater. It is the organizational discipline that determines whether AI systems serve the purposes for which they were deployed, treat affected individuals fairly, and remain within the risk tolerances that the organization has committed to uphold.

The manager who finishes this book should leave with several durable conclusions. First, AI risk is not a technology problem that can be delegated to a data science team and forgotten. It is a management responsibility that requires the same organizational infrastructure as any other significant operational risk: defined controls, continuous monitoring, independent audit, and escalation paths that reach senior leadership. Second, governance that is not operationalized is not governance. Policies that exist but are not enforced, controls that are designed but not tested, and monitoring systems that alert but trigger no response are not governance assets. They are liability artifacts that suggest awareness without action.

Third, the regulatory and legal environment for AI will continue to tighten. The current period is one of norm formation: principles articulated in frameworks like the Blueprint, the NIST AI RMF, and the EU AI Act are becoming the evidentiary standard against which organizations will be judged in court and before regulators. Organizations that build governance programs aligned to these frameworks now will find that incremental compliance with future binding requirements is manageable. Those who wait for enforcement will build under pressure with inadequate foundations.

Fourth, the organizational capability required for AI governance is buildable with deliberate effort. None of the controls, artifacts, or processes described in this book requires specialized expertise that is unavailable to a reasonably staffed enterprise IT organization. What is required is leadership commitment to the decision that governance is a first-class organizational priority, not a cost center to be minimized. That decision is the manager's to make, and this book has been an attempt to give the manager the knowledge and the tools to make it confidently.

Fifth, governance quality compounds. Organizations that build strong foundations in their first year find that each subsequent year of governance effort produces more value. Because the evidence portfolio is deeper, the monitoring data has a longer history, the team has developed institutional knowledge, and the processes are increasingly embedded in organizational habit. The early investment is the highest. The value accumulates. Managers who start now, even imperfectly, are ahead of managers who wait for the perfect framework, the perfect tooling, or the perfect mandate.

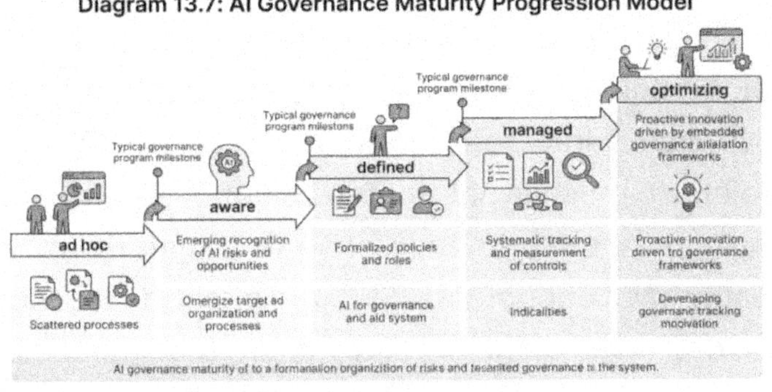

Diagram 13.7: AI Governance Maturity Progression Model

13.9 Manager's Closing Checklist

Across the full program: maintain a current AI system inventory covering all production and near-production systems with ownership, purpose, risk classification, and documentation status. Hold all system owners accountable for maintaining current model cards and contributing to the portfolio risk register. Consistently operate the three-tier governance cadence, producing written action records from every meeting. Conduct annual independent reviews for high-stakes systems and act on findings within committed timelines. Maintain the regulatory landscape watch, assign someone to monitor for new AI-relevant requirements, and update the governance framework accordingly.

Culture and capability: make incident reporting psychologically safe and operationally straightforward. Invest in ongoing training so that AI system owners understand their governance responsibilities and the rationale behind them. When governance requirements conflict with deployment timelines, escalate; do not quietly deprioritize the governance requirement. Build governance into the definition of "done" for every AI project so that it is not treated as optional overhead. The measure of a governance program's maturity is not the quality of its documents. Still, the quality of its decisions and decisions of that quality require the habits that only consistent practice builds.

Every AI governance program starts somewhere. The organization in this book's opening chapter, the one that discovered its governance vacuum when a regulator asked a question, completed its remediation and went on to build a program that became a competitive differentiator in its market. Clients who cared about responsible AI chose them over competitors who could not demonstrate equivalent governance discipline. The program that started as a regulatory response became a business asset. That trajectory is available to any organization that decides governance is worth doing well. The tools are in this book. The decision is yours.

14 THE GOVERNED ENTERPRISE

Governance does not announce itself. It operates quietly in the background of every AI deployment decision in the risk assessment that precedes a model's approval for production, in the audit trail that records who authorized a change and why, in the policy that specifies how training data may and may not be used, in the escalation path that routes a bias complaint to the right committee rather than leaving it unaddressed in an engineer's inbox. When governance is working, organizations rarely notice it. When it is absent, the failures are unmistakable: regulatory findings, reputational damage, operational disruptions that cascade in ways nobody anticipated.

This book has moved through thirteen chapters to build the architecture of a governed enterprise, one that deploys AI not simply because it can but because it has established the structures, processes, and accountabilities that make responsible deployment possible. This closing chapter draws the architecture's threads together, identifies the themes that run beneath the chapter structure, and offers a forward-looking perspective on where AI governance is headed and what it will ask of managers in the years ahead.

14.1 The Persistence of Human Judgment

The first theme is the one that runs most insistently through every chapter: governance is not an automated function. It is an exercise of human judgment, applied systematically. Risk registers require someone to decide which risks to include and how to rate their likelihood and impact. Policy frameworks require someone to determine which uses are acceptable and what exceptions are warranted. Audit programs require someone to interpret findings and determine which warrant escalation. Lifecycle checkpoints require someone to decide whether a model is ready for deployment or requires additional validation.

This is not a limitation of current technology. It is a permanent feature of governance. The value of governance frameworks, templates, and tools is that they structure judgment—they ensure it is applied consistently, documented, and visible to oversight, not that they replace it. The manager who understands this will build governance programs that invest in the capacity for judgment: clear decision rights, well-designed committees, trained reviewers, and escalation paths that bring the right expertise to bear when questions are close calls.

Organizations that attempt to automate their way out of governance will find that they have automated their compliance artifacts while leaving their governance substance hollow. Regulators are increasingly sophisticated about this distinction. An audit trail that records automated approvals without evidence of substantive human review does not satisfy the oversight requirements that AI governance frameworks are designed to fulfill.

14.2   The Integration Imperative

The second theme is integration. AI governance that operates as a parallel structure separate from existing IT governance, risk management, compliance, and legal functions will consume resources, create friction, and ultimately fail to embed governance in the decisions that matter most. The chapters on IT governance alignment, policy architecture, and the AI development lifecycle all point to the same conclusion: governance is effective when integrated into the workflows that produce and operate AI systems, not when it audits those workflows after the fact.

This integration is organizationally demanding. It requires IT governance and AI governance to develop shared frameworks rather than separate ones. It requires engineers to internalize governance checkpoints as part of their development practice rather than as compliance gates that slow releases. It requires legal and compliance functions to provide guidance early enough in the development lifecycle to shape decisions rather than review them retrospectively. None of this happens without deliberate organizational design—governance structures, incentives, and communication patterns that bring these functions into genuine coordination.

The payoff is proportionate to the investment. Organizations that achieve genuine integration deploy AI faster because governance reviews are embedded in development rather than appended to it. They satisfy regulators more efficiently because their documentation practices produce evidence as a natural byproduct of operations rather than requiring retrospective reconstruction. They respond to incidents more effectively because their governance structures include clear escalation paths, defined roles, and practiced postmortem processes. Integration is not merely a best practice—it is the difference between a governance program that works and one that merely exists.

14.3   The Maturity Trajectory

The third theme is maturity. AI governance is not a binary state, present or absent, but a capability that organizations develop over time. The 90-day implementation roadmap in Chapter 13 describes the initial phase of that development: standing up the core structures, establishing foundational policies, and activating basic monitoring and audit practices. That is the beginning of a trajectory, not the destination.

Governance maturity in the context of AI has several dimensions. Structural maturity: moving from ad hoc governance decisions to formal committees, clear decision rights, and documented escalation paths. Process maturity: moving from retrospective compliance reviews to proactive risk assessment integrated into every stage of the AI lifecycle. Technical maturity: moving from manual audit practices to automated monitoring for model drift, fairness degradation, and security anomalies. Organizational maturity: moving from governance as a function performed by a dedicated team to governance as a disposition shared across the AI-producing and AI-operating workforce.

Progress along each of these dimensions requires measurement. Governance programs that do not track their own performance cannot demonstrate improvement or identify the points where investment is most needed. The metrics referenced throughout this book, policy coverage, risk register currency, audit finding resolution times, training completion rates, and incident detection and response times, are not bureaucratic artifacts. They are the instrumentation that allows governance leaders to manage their programs rather than maintain them.

14.4   The Regulatory Horizon

The fourth theme is the regulatory environment, which is certain to become more demanding. The current landscape, a patchwork of sector-specific guidance, national frameworks, and emerging statutes, is the beginning of a regulatory trajectory, not its mature state. The European Union's AI Act will generate compliance requirements that extend beyond EU-based organizations to any entity whose AI systems affect EU residents. National AI legislation in the United States is advancing at both the federal and state levels, with sector-specific agencies moving on parallel tracks. International standards bodies are developing AI governance standards that will shape procurement requirements, contractual expectations, and audit frameworks across industries.

The organizations best positioned for this regulatory trajectory are not those that respond to each new requirement as it emerges, but those that have built governance programs with the structural capacity to adapt. A governance framework designed for auditability, with clear documentation practices, maintained audit trails, and regular independent reviews, can incorporate new requirements at a lower marginal cost than a framework built reactively to specific mandates. The investment in governance architecture pays regulatory dividends over time.

The manager's role in this environment is to maintain situational awareness of the regulatory horizon not by becoming a legal expert but by maintaining relationships with legal counsel, participating in industry working groups, and establishing monitoring processes that surface relevant regulatory developments before they become compliance deadlines.

14.5   The Manager as Governance Steward

The fifth theme is the manager's own role, which is the subject of this series and the thread that holds these thirteen chapters together. Governance stewardship is the active, ongoing exercise of accountability over AI systems. It is distinct from technical management (ensuring systems perform correctly), project management (delivering systems on schedule and within budget), and compliance management (satisfying regulatory requirements). It encompasses all of these while adding the dimension of institutional accountability: the obligation to ensure that the organization can account to its stakeholders, regulators, customers, employees, partners, and the public for its AI decisions.

Governance stewardship is exercised through specific practices: decisions made by risk committees, standards set for policy documentation, the emphasis placed on training and awareness, the seriousness with which audit findings are addressed, and the investment in assurance programs that go beyond minimum compliance requirements. It is also exercised within organizational culture through norms that determine whether governance is treated as a genuine obligation or a formality, whether risk identification is rewarded or suppressed, and whether escalation paths are used or avoided.

The manager who governs AI well is not necessarily the one with the most sophisticated technical understanding. They are the ones with the clearest understanding of their accountability, the most disciplined approach to establishing and maintaining governance structures, and the deepest commitment to building an organization that can answer honestly for the systems it deploys.

14.6 Practical Next Steps

For the manager who has read this book and is now considering next actions, the path forward has a clear initial structure.

Assess your current state. Before building or rebuilding a governance program, understand what currently exists. Map the AI systems in your portfolio. Identify which have documented risk assessments, which have policy coverage, which have monitoring in place, and which do not. The gap between the current state and the governed state is the program you need to build.

Prioritize by risk. Not all AI systems require the same level of governance. Systems that make or significantly influence decisions affecting individuals' hiring, lending, healthcare, benefits, and law enforcement require more rigorous governance than internal operational tools. Prioritize your governance investment based on each system's risk profile, not on the organizational visibility of its sponsors.

Build the foundation first. The 90-day roadmap in Chapter 13 describes the foundational work that all subsequent governance capabilities depend on: establishing the committee structure, standing up the risk register, activating the policy framework, and initiating the training program. Resist the temptation to build sophisticated monitoring systems before the foundational accountability structures are in place. Automation amplifies governance programs; it cannot substitute for them.

Measure and report. Establish governance metrics from the beginning and report on them consistently. The act of measuring governance performance changes the program's relationship to organizational leadership: it makes governance visible as an operational function rather than a background compliance activity, and it creates the accountability dynamic that sustains investment in the program over time.

Stay connected to the field. AI governance is a rapidly evolving discipline. The frameworks, standards, and regulatory requirements in effect today will be revised, supplemented, and superseded. The manager who maintains connections to professional communities, regulatory guidance, and peer organizations will be positioned to adapt with lower friction than one who treats this book as a finished map rather than a starting point.

The governed enterprise is not a destination; it is a posture, maintained through continuous deliberate effort. The effort is worth making. The organizations that govern AI with rigor and integrity will be the ones that deploy it with confidence, sustain it with accountability, and earn the trust of the stakeholders whose lives those systems touch.

That is the work. These chapters have provided the architecture. The stewardship is yours.

15 References

Algorithmic Accountability Office. AI Audit Framework: Standards for Organizational Assurance. Algorithmic Accountability Office, 2025. (CH12)

Association for Computing Machinery (ACM). ACM Code of Ethics and Professional Conduct: AI Supplement. ACM, 2024. (CH2, CH3)

Center for Security and Emerging Technology (CSET). Mapping the AI Governance Landscape: April 2026 Update. CSET, 2026. (CH3, CH12)

CISA. A CISA Roadmap for Artificial Intelligence 2023–2024. Cybersecurity and Infrastructure Security Agency, Nov. 2023. (CH5, CH11)

Deloitte. The State of AI in the Enterprise 2026. Deloitte Insights, 2026. (CH9, CH13)

European Commission. Proposal for a Regulation Laying Down Harmonized Rules on Artificial Intelligence (AI Act). European Commission, 2024. (CH4, CH10)

ENISA. Post-Quantum Cryptography: Preparing for the Transition. European Union Agency for Cybersecurity, 2024. (CH7, CH9)

Global Partnership on AI (GPAI). Responsible AI Implementation Guide for Public Sector Organizations. GPAI, 2025. (CH2, CH5)

Harvard Kennedy School, Belfer Center. AI Governance Playbook for Public Sector Leaders. Belfer Center, 2024. (CH3, CH5)

IEEE Standards Association. Guidance on Cryptographic Agility and Post-Quantum Migration. IEEE, 2025. (CH7)

IEEE Security & Privacy. Special Issue: Post-Quantum Cryptography and Enterprise Migration. IEEE, 2024. (CH3, CH8)

IETF. RFC: Post-Quantum Key Exchange and Hybrid TLS Extensions. Internet Engineering Task Force, 2025. (CH6)

IAPP. AI Governance Vendor Report 2026. International Association of Privacy Professionals, 2026. (CH3, CH7)

MIT AI Policy Lab. AI Risk Management: Operationalizing the NIST AI RMF in Practice. MIT, 2024. (CH3, CH12)

MIT Technology Review Insights. AI Assurance and Audit: Emerging Practices for Enterprises. MIT TR Insights, 2025. (CH12)

National Institute of Standards and Technology (NIST). Artificial Intelligence Risk Management

Framework (AI RMF) 1.0. NIST, Jan. 2023. (CH3, CH12)

NIST. *AI RMF Playbook: Implementing Trustworthy AI Practices*. NIST, 2023. (CH3, CH5)

NIST. *NIST-AI-600-1: Generative AI Profile for the AI RMF*. NIST, July 2024. (CH11)

NIST. *Concept Note: AI RMF Profile on Trustworthy AI in Critical Infrastructure*. NIST, Apr. 2026. (CH3, CH5)

NSA. *Commercial National Security Algorithm (CNSA) Suite 2.0 Guidance and Transition Roadmap*. National Security Agency, 2024. (CH4, CH7)

OECD. *Progress in Implementing the European Union Coordinated Plan on AI (Volume 1)*. OECD.AI, Nov. 2025. (CH4, CH10)

Oxford Internet Institute. *Red Teaming and Adversarial Testing for AI Systems: Methods and Metrics*. OII, 2025. (CH11, CH12)

Oxford Quantum Group. "Assessing Logical Qubit Roadmaps and Cryptographic Relevance." *Quantum Information Journal*, 2025. (CH7)

Pfleeger, Charles P., and Shari Lawrence Pfleeger. *Cryptographic Agility: Principles and Practice*. Wiley, 2025. (CH7)

RAND Corporation. *Governing Generative AI: Policy Options and Operational Controls*. RAND, 2024. (CH3, CH11)

Schneier, Bruce, et al. "Operationalizing Post-Quantum Cryptography in Large-Scale Systems." Communications of the ACM, 2025. (CH7, CH8)

Stanford HAI (Human-Centered AI). Long-Term Data Confidentiality and Harvest-Now, Decrypt-Later Threats. Stanford HAI White Paper, 2024. (CH1, CH7)

The White House, OSTP. Blueprint for an AI Bill of Rights: Executive Summary and Implementation Guidance. Office of Science and Technology Policy, 2024. (CH4, CH10)

US Department of Commerce. Guidance on Federal Adoption of Post-Quantum Standards. U.S. Dept. of Commerce, 2024. (CH4, CH8)

US Office of Management and Budget (OMB). Memorandum: Federal Agency Planning for Post-Quantum Cryptography. OMB, 2025. (CH8, CH9)

Van Meter, Rodney, and K. Brown. "Hardware Security Modules and PQC Acceleration: An Evaluation." Journal of Applied Cryptography, 2025. (CH7)

World Economic Forum. Quantum Computing and Cybersecurity: Strategic Implications for Enterprises. WEF Report, 2024. (CH1, CH9)

World Health Organization (WHO). Ethical and Governance Considerations for AI in Health: A WHO Guidance Document. WHO, 2024. (CH2, CH9)

Zhang, L., et al. "Prompt Governance and Hallucination Mitigation for Large Language Models." IEEE Transactions on AI Ethics and Safety, 2025. (CH11)